EXPERIMENTS

IN

BIOCHEMICAL

RESEARCH

TECHNIQUES

ROBERT W. COWGILL

- 57-2970

Department of Biochemistry
University of Colorado School of Medicine

ARTHUR B. PARDEE

Department of Biochemistry and
Virus Laboratory
University of California

EXPERIMENTS

IN

BIOCHEMICAL

RESEARCH

TECHNIQUES

NEW YORK · JOHN WILEY & SONS, INC.

London · Chapman & Hall, Ltd.

PREFACE

This book contains a selection of experiments that are intended to illustrate some of the major research techniques of modern biochemistry. For the past few years, these experiments have constituted the elective and required laboratory material for two consecutive one-semester courses attended by first-year graduate students in biochemistry and related fields at the University of California (Berkeley). For these courses, we have selected a limited number of experiments of moderate complexity in order to teach the techniques within a reasonable time. Also, we have tried to develop a "research attitude" in our students by presenting the experiments in such a way that the students must acquire some of the details from their own reasoning and ingenuity or from references to the literature.

We must justify the need for a course in advanced biochemical techniques, for it may reasonably be asked whether graduate students could not spend their time more profitably in other pursuits, for example, in actual research. It is true that the student can acquire the techniques presented in this book from another research worker when the need arises; or he may even work out the methods for himself if necessary. However, these alternative ways of learning are themselves time-consuming and inefficient. What is more important, one tends to apply to a research problem those methods with which one is familiar and hesitates to take the time to learn new methods; hence, quite useful techniques may never be applied because of an "inertia of ignorance." Therefore, we feel

that it is worth our students' time to become acquainted, through a laboratory course, with a variety of the techniques that they will need most in
their research. This manual was written to provide the necessary experiments, since no published compilation was available.

The question of which techniques are the most important also must be
considered. It is far too much to expect that everyone should hold the same
ideas on this subject. The most we can hope is that our selection will satisfy a few people, will be acceptable in part to many others, and therefore
will provide a selection of useful experiments. A number of alternative
experiments have been provided to permit a wider choice. We have omitted
experiments of the kind given in basic biochemistry courses because they
should have been performed earlier by the students for whom this manual
is intended. A number of manuals that contain these basic experiments are
available. Certain highly specialized techniques that require elaborate
equipment (such as ultracentrifugation and boundary electrophoresis),
microbiological and hormone assays, and nutritional experiments have
been omitted. Experiments based extensively on organic synthesis or
degradation were not included. To recapitulate, selection was necessary
in order to keep the amount of work expected of the students within reasonable limits of time and effort, and our choice has been based primarily
upon our interests and experience. We shall be most grateful for suggestions regarding either the individual experiments or the selection of experiments in this manual.

We should like to express our gratitude to Dr. Willis H. Reisen who
assembled certain of the experiments in Sections I and III of this book. We
also are indebted to Dr. A. K. Balls of Purdue University, Dr. A. L. Marr
of the University of California at Davis, Dr. L. K. Noda and Dr. F. W.
Strong of the University of Wisconsin, and Dr. Harland Wood of Western
Reserve University for providing us with experiments used in their courses.
Our greatest debt is to our students and our teaching assistants who tested
these experiments and who suggested many improvements.

Denver, Colorado   ROBERT W. COWGILL

Berkeley, California   ARTHUR B. PARDEE

May 1, 1957

CONTENTS

GENERAL INTRODUCTION

The experiments in this manual are intended for advanced students and many of them are designed as problems in research, not as routine laboratory exercises. Consequently, the procedures described are in some cases little more than guideposts to you, the researcher. You will be expected to give some thought to modifications of experimental procedures, and to supplement the text instructions by reading the original literature for more details.

Two rules for your safety are to be stressed.

a. No student is to work in the laboratory when he is alone.

b. Students are not to smoke nor have flames near apparatus unless sure the contents are non-flammable.

Every experiment must be clearly described in your notebook. The essential characteristic of a good laboratory notebook is that it be sufficiently detailed and neat so that someone else (or you yourself) could repeat the experiment and obtain comparable results. The following material should be included in your notebook.

1. Date of the experiment.

2. Title: a quick reference to the subject of the experiment.

3. Purpose: a clear statement of the questions the experiment is designed to answer.

4. References: a citation of essential literature.

5. Procedure: a description clear enough for duplication of the experiment at a later date. You need not copy material from other sources un-

1

less major changes have been made, but a reference should be provided.

6. Results: a direct record of data (which subsequently may be reorganized in the form of tables or charts). If it is necessary to use results obtained by other students, the source of the data should be acknowledged.

7. Discussion and conclusions: a brief summary of your discoveries and their relation to results obtained by others. Such a summary is very useful later, when you wish to recall quickly what the experiment showed.

You will find that each experiment in this manual is arranged under certain headings (Objectives, Principles, etc.). The Objectives of each experiment are stated first. Before commencing your laboratory work you should understand the purpose of a particular experiment and what you are expected to learn from it.

The Principles, given next in some experiments, are brief summaries of background material and other information useful for the experiment. The major references given in this section are especially important and you are expected to read them before beginning the actual experimental work.

The list of Principal Equipment and Supplies contains only the less common pieces of apparatus, solutions, and chemicals you will need for the experiment. It is your responsibility, however, to see that everything required is ready when you need it. Be certain, for example, that such things as sterile media, cold solutions, bacterial cultures, and warm-water baths are prepared ahead of time.

The Procedure should be read completely before the experimental work is commenced. The point to be stressed is that you should know what you are going to do and why, before you come to the laboratory. The following general suggestions for working in the laboratory are offered at this point.

a. Perform odd jobs such as cleaning equipment or calculating results in the intervals that occur during an experiment between major operations. This sort of planned procedure is a key to efficient laboratory work; you should never find yourself with nothing to do.

b. Use clean equipment and keep your desk neat while you work.

c. You will not become so tired while you work if you sit down whenever possible.

d. When you have a question, try to answer it for yourself before you see the instructor. Ask yourself what experiments you would have to do to obtain an answer; you may find that you already have the information required to answer the question.

e. Try to develop some original ideas about the experimental procedures, and try to plan more interesting experiments. Check with the instructor before attempting any major change, however, to be sure that it is feasible and that equipment is available. There are often good reasons for the specific details in the given procedures.

The section on Treatment of Data is deliberately varied from one experiment to another to help you to develop some skill in reporting scientific work. Reports are oral or written, formal or informal. Formal, written reports are usually the most difficult to prepare but books are available to help you in this task (1, 2). Brief and clear directions on how to revise your report for greater clarity and better grammar also have been published (3).

The Questions section of each experiment is designed to bring to your attention concepts and procedures that were not stressed in the text of the experiment.

The References listed at the end of each experiment have been kept to a minimum number. The major references, numbered in underlined type, are articles which will help you to acquire a clear understanding of the important points of each experiment. At least the portions of these references that are applicable to the experiments should be read. Supplementary references generally contain information of a more specialized nature.

It is most important for a scientist to read the literature. In actual research, much time in the laboratory can be saved by reading pertinent articles both before starting a new problem and again after work has begun. During these experiments, too, you should read the essential material in the literature. A second reason for reading original articles and reviews is to broaden your background and acquire new concepts. The importance of this reading cannot be overestimated. You should develop a habit of skimming through titles and summaries of the current articles in several Journals. It is impossible to read even a fraction of the total that appears, and you should read an entire article only when it becomes essential for your research.

Biochemical literature appears in different forms, and serves different purposes.

1. Biochemistry texts are useful to gain a preliminary, very general view of a topic. The treatment of specific problems is too brief for use in research and the texts may be out of date by five years or more in regard to specific points.

2. Specialized reference books, such as The Enzymes (4), will provide summaries of basic knowledge and are especially useful for references to earlier papers of importance.

3. Once you have gained a general idea of your topic, some recent specialized review can be extremely helpful. Annual Reviews (of Biochemistry, Microbiology, Physiology, etc.), Advances (in Enzymology, Carbohydrate Chemistry, etc.), or similar review series are examples. Start with the most recent volume and work backwards; note references to pertinent original papers and reviews.

4. Abstract Journals (Chemical Abstracts, Biological Abstracts, etc.) are useful for locating all published work except that appearing in the past year. However, it is often easier to find important papers in reviews than in abstracts.

5. The current Journals must be searched for articles published within the past year. The references listed in review articles should give you an indication of the Journals that are likely to carry pertinent articles. In reading articles, start with the recent ones, for they will provide references to earlier work.

6. Abstracts of recent scientific meetings (Federation Proceedings, American Chemical Society Meetings, etc.) contain brief descriptions of very recent, unpublished work, and these indicate what work is likely to be going on at present.

Some references may be made to books dealing with the way biochemical research is actually performed. A most worthwhile book by Beveridge (5) points out, among other things, that keen observation in the laboratory and a questioning attitude toward unexplained results lead to many important discoveries, and also that an initial step in research often arises from intuition and inspiration of an artistic sort. Reason later provides a check. A most interesting book on the scientific life of Pasteur (6) is valuable reading for students looking toward a career in science.

References

1. Trelease, S. F. 1947. The Scientific Paper. Williams and Wilkins. Baltimore, Md.
2. Ulman, J. N., Jr. 1952. Technical Reporting. Henry Holt and Co. New York.
3. Keck, W. M. 1956. Checklist for Thesis Revision. Industrial Science and Engineering, 3, 19–22.
4. Sumner, J. B., and Myrback, K. 1950. The Enzymes. Academic Press. New York.
5. Beveridge, W. I. B. 1952. The Art of Scientific Investigation. W. W. Norton and Co. New York.
6. Duclaux, E. 1920. Pasteur; The History of a Mind (translated by E. F. Smith and F. Hedges). W. B. Saunders Co. Philadelphia, Pa.

PHYSICAL CHEMICAL METHODS FOR SEPARATION AND IDENTIFICATION OF BIOLOGICALLY IMPORTANT COMPOUNDS

Frequently an investigator in the field of biochemistry is faced with the problem of separation and identification of compounds in a mixture, the components of which are structurally similar and, therefore, possess very similar physical and chemical properties (mixtures of nucleotides, of fatty acids, of proteins or their degradation products, etc.). Often only a partial separation of structurally similar compounds is effected by a single performance of a given separation process, such as a single extraction or a single adsorption on charcoal. Where separation does depend upon subtle differences in physical or chemical properties, a repetition of the basic separation process is required in order to approach complete separation. This may take the form of a cascade-like repetition of the unit process; thus countercurrent distribution may be viewed as a systematic, repetitive process of distribution between two liquid phases. Or, the repetition may take the form of a continuous, non-differentiated process in which unit steps are no longer distinguishable; thus partition column chromatography may be viewed as a process of continuous distribution and redistribution of solutes between two phases throughout the column. Because of their extreme effectiveness, many of the methods of separation and identification described in this section are of the continuous, non-differentiated type (Table I).

This section will be limited to certain methods of wide applicability for which apparatus is likely to be available or readily procurable. The section does not include such highly useful methods for the study of large

5

TABLE I. TECHNIQUES FOR SEPARATION BY PHASE DISTRIBUTION

Type of Separation Process	Phase Pairs Involved		
	Vapor-Liquid	Liquid-Liquid	Liquid-Solid
Batchwise	Simple distillation	Extraction	Decolorization by adsorption
Cascade	Bubble-cap distillation	Countercurrent distribution	Systematic, repeated adsorption
Continuous, non-differentiated	Packed column distillation	Liquid-liquid partition chromatography	Adsorption chromatography
	Gas-liquid partition chromatography	Paper chromatography	Ion exchange chromatography

molecules as the ultracentrifuge, boundary electrophoresis, light scattering, etc., because of the elaborate equipment required. Also, such rather simple separation procedures as the commonly used isolation procedures for proteins, fats, nucleic acids, etc., are not considered in this section.

A. Distillation at Low Pressures

Distillation is the oldest of all techniques described in this section and is, in fact, one of the oldest of all chemical processes. It may be employed for purification of a compound or for proximate analysis of a mixture. Amounts of material from about two grams to several kilograms may be handled in suitably designed laboratory apparatus. The components of the typical distillation system for pressures down to about 1 mm of Hg are shown in Figure 1. Descriptions of the multitude of types, capacities, and pressure ranges of the various components of the distillation system are set forth in refs. (1) and (2). In assembly of such a system, one should first establish the desired capacity and operating pressure ranges, then consult the above references for the best combination of components for the given system.

Although low-pressure distillation encompasses all distillations at less than atmospheric pressure, most fractional distillations at low pressure are conducted in the range of 1–20 mm of Hg. One advantage of low pressure for distillation of biologically important compounds is the decreased thermal decomposition at the lower distillation temperature, and a possible second advantage is a change in relative volatility of components; for example, ethanol and water form constant boiling (azeotrophic) mixtures at atmospheric pressure but do not do so at pressures below 70 mm of Hg and can be readily separated at low pressures.

Certain biological materials such as the oil-soluble vitamins require

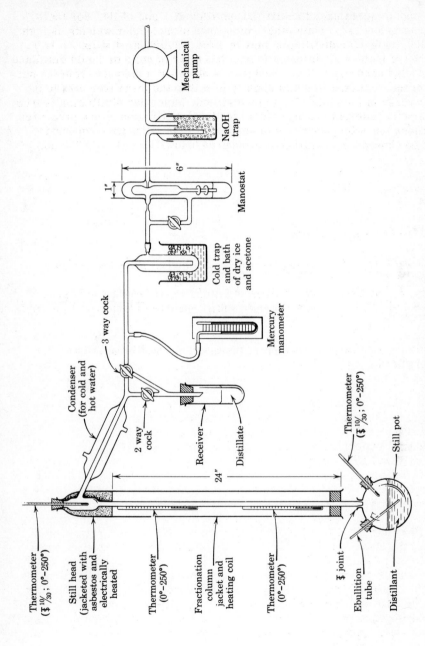

Figure 1. Apparatus for distillation at low pressures.

very high temperatures for distillation, even at 1 mm of Hg. For these compounds and for certain other compounds of molecular weights as high as 1200, molecular distillation may be used. Molecular distillation is a process of molecular diffusion from a hot layer of solid or liquid distilland to a cooled condenser. The condenser is placed at a distance from the surface of the distilland which is slightly less than the mean free path in the vapor phase of the molecules to be distilled. Molecular distillation is thus a distinctly different process from fractional distillation at low pressures; it utilizes much the same type of vacuum system, although components must be chosen for operation at pressures below 10μ $(1\mu = 0.001$ mm of Hg).

References*

1. Carney, T. P. 1949. Laboratory Fractional Distillation. Macmillan Co. New York.
2. Rose, A., and Rose, E. 1951. Theory of Distillation. In Technique of Organic Chemistry, IV, 1–174. A. Weissberger, editor. Interscience Publishers. New York.

---

*Major references are designated throughout this book by underlined reference numbers.

EXPERIMENT 1. LOW PRESSURE DISTILLATION OF METHYL
ESTERS OF FATTY ACIDS (4 periods)

OBJECTIVE

This experiment will provide experience in the operations of vacuum distillation and an appreciation of the characteristics of various types of apparatus.

PRINCIPAL EQUIPMENT AND SUPPLIES

Twenty-four-inch Podbielniak column with partial reflux head (ref. [1], p. 237), and/or
Thirty-inch Vigreux column with variable take-off head (ref. [1], p. 245), or other columns of comparable efficiencies at low pressures
Mechanical pump
Auxiliary distillation equipment shown in Figure 1
Abbé refractometer and circulating water bath at 45°
Steam bath
Sample of coconut oil of unknown composition
Absolute methanol
Saturated sodium chloride
0.5 N alcoholic KOH
0.5 N HCl
Bromphenol blue (0.04%)
Phenolphthalein in alcohol (1%)
Methyl orange (0.04%)

PROCEDURE

You should be familiar with the theory of fractional distillation and the operation of the equipment before you attempt this experiment. It is recommended that the references listed at the end of this experiment be consulted; in particular, ref. (1).

1. *Formation of Methyl Esters of the Fatty Acids* (2)

To 100 gm of coconut oil in a 500 ml round-bottom flask are added 250 ml of absolute methyl alcohol and 5 ml of concentrated H_2SO_4. After the mixture has been refluxed on the steam bath for 20–24 hr an equal volume of saturated sodium chloride solution is added, and the aqueous phase of the mixture is neutralized to methyl orange with powdered sodium carbonate. The layer of mixed methyl esters is separated, washed twice with water, and dried by heating to 110° in a beaker for 30 min.

2. *Fractional Distillation of the Methyl Esters*

The oily liquid is transferred to a three-neck, 250 ml, round-bottom flask with standard taper fittings. The center neck joins the distillation column while into one side neck is fitted a thermometer. Into the other side neck is fitted an ebulition tube in the form of a capillary tube drawn from 6 − 8 mm tubing. The capillary should be of such bore that you can blow bubbles from it by mouth into acetone but not into water. The still pot is immersed in an electrically heated oil bath (see Figure 1). Distillation is started at about 5 mm of Hg; a manostat is used to maintain constant pressure. Adjust heaters for pot and column until the column is operating very near its capacity, i.e., just below the flooding stage. At the start the column is operated at total reflux for several minutes to insure equilibration; then the distillate is taken off at a reflux ratio of at least 5 : 1.

Flooding or bumping should not occur during the fractionation, and if any such irregularity does occur the column should be placed on total reflux until the difficulty is overcome. The boiling points of the various esters are plotted against pressure in Figure 2. Distillation in the range

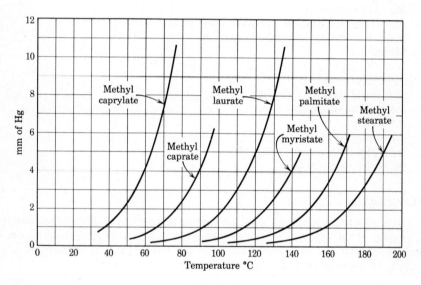

Figure 2. Boiling points of methyl esters of fatty acids.

2 − 5 mm of Hg is convenient for the separation of esters in your sample and fractionation should certainly not be attempted below 1 mm of Hg. Follow the course of the distillation by a plot of head temperature versus cumulative weight of distillate, and divide the distillate into fractions of about 3 − 5° range in boiling point. Do not continue the distillation beyond a pot temperature of 250°. A representative data sheet for the preservation of data collected during the experiment is given in Table II. A period of 6 − 8 hr may be required for the complete distillation, but it is advisable to complete the distillation without interruption if this is at all possible.

TABLE II. REPRESENTATIVE VACUUM DISTILLATION
RECORD SHEET

Sample: . . . .   Sample Weight:. . . . .

Type of Column:. . . . .   Date:. . . . .

Fraction No.						
Gross						
Tare						
Net weight						
Bath temp.						
Bath heater rheostat						
Pot temp.						
Column temp.						
Column heater rheostat						
Head temp.						
Head heater rheostat						
Pressure						
Time						
n_D^{45}						
Comments						

Residue (approximate weight)

3. *Determination of Refractive Index* (3)

Measurements of refractive index are among the easiest and most accurate measurements of physical chemistry. Determinations can be made easily to two parts in ten thousand, but careful control of temperature is essential for accurate work. The refractive index is determined conveniently with a refractometer which is direct reading and which includes a compensator so that the observed values with white light correspond to those obtained with the wave length of the D line of sodium. The following directions are for the use of an Abbé refractometer of the above type.

An auxiliary prism below the refracting prism serves both to illuminate and to confine the sample. The auxiliary prism is hinged so that its ground glass surface can be clamped nearly in contact with the refractive prism.

a. The surfaces of the prisms must be protected carefully; clean the surfaces with alcohol and wipe them with soft lens paper before and after each reading. Leave no lint on the surfaces.

b. Apply a coating of the methyl ester fraction by means of a flame-polished stirring rod to the ground surface of the auxiliary prism. Be sure that the liquid is evenly spread and that air bubbles are not entrapped. Lock the two prisms together whereby a thin film (0.1 mm) of liquid is held against the upper prism.

c. Adjust the mirror to give maximum illumination of the field.

d. Adjust the prisms by movement of the alidade (scale index arm which is attached to the prisms) until the critical boundary is in the middle of the field.

e. Achromatize the boundary by rotation of the compensator in the eyepiece.

f. Align the critical boundary with the intersection of the cross hairs in the eyepiece. Read the refractive index from the scale and vernier on the alidade.

g. A second alignment should be made with the compensator in the other achromatizing position. The two index readings will probably differ. However, the mean value is free from the particular instrumental error which gives rise to this effect.

h. After a measurement, clean the prisms at once, because the sample may dissolve the prism mounting cement.

The refractive index is recorded as n_D^t in which t is the temperature of the sample and D refers to the wave length of light, i.e., the D line of sodium. The refractive indexes of the methyl esters of fatty acids present in coconut oil are given in Table III.

TABLE III.   REFRACTIVE INDEXES OF METHYL ESTERS (5)

Ester	Mol Wt	n_D^{45}
Me caprylate	158	1.4069
Me caprate	186	1.4161
Me laurate	214	1.4220
Me myristate	242	1.4281
Me palmitate	270	1.4317
Me stearate	298	1.4346
Me oleate	296	1.4520

4. *Determination of the Saponification Equivalent* (6)

Determine the saponification equivalent of at least two of your best fractions and compare the results with theoretical values. Weigh about

1.0 gm of the ester into a 50 ml Erlenmeyer flask and add approximately 15 ml of 0.5 N alcoholic KOH. Attach the flask to an air condenser with a soda-lime protection tube at the top and reflux the mixture gently for 2 hr on a water bath or steam plate. Cool the contents of the flask and add 3 drops of phenolphthalein indicator. Add 0.5 N HCl until the pink color just disappears (you do not need to measure the amount of HCl added). Then, add 1 drop of bromphenol blue indicator and 3 ml of benzene, and again titrate with 0.5 N HCl but this time measure the HCl added. Titrate to a green color which does not change upon swirling the contents to mix the two phases. Be sure to let the phases separate well before judging the color of the aqueous solution. The saponification equivalent = 1000/(normality of HCl × titer to second end point).

TREATMENT OF DATA

Tabulate the distillation data as shown in Table II. Prepare graphs by plotting the weight of distillate obtained against head temperatures and refractive indexes. Estimate the per cent (by weight) composition of the unknown sample. Preserve the fractions which you consider to be pure and summarize your evidence for their identity and purity.

QUESTIONS

1. Why is fractional distillation not done at pressures much below 1 mm of Hg?
2. Why is a manostat used to regulate pressure rather than a controlled air leak?
3. Account for the presence and specific location of the traps, gage, and manostat in Figure 1.
4. What purpose does the benzene phase serve in the titration of the saponified fatty acids?

References

1. Cason, J., and Rapoport, H. 1950. Vacuum Distillation. In Laboratory Text in Organic Chemistry, 232–59. Prentice Hall. New York.
2. Sauer, J. C., Hain, B. E., and Boutwell, P. W. 1940. Methyl Myristate and Methyl Palmitate and the Corresponding Acids. Org. Syntheses, 20, 67–70.
3. Bauer, N., and Fajans, K. 1949. Refractometry. In Technique of Organic Chemistry, 2nd ed., Vol. I, Pt. II. A. Weissberger, editor. Interscience Publishers. New York.
4. Scott, T. A., Jr., MacMillan, D., and Melvin, E. H. 1952. Vapor Pressure and Distillation of Methyl Esters of Some Fatty Acids. Ind. Eng. Chem., 44, 172–5.
5. Wyman, F. W., and Barkenbus, C. 1940. Methyl Esters of the Higher Fatty Acids. Ind. Eng. Chem., Anal. Ed., 12, 658–61.
6. Rieman, W. 1943. Determination of the Saponification Number of Fats and Oils. Ind. Eng. Chem., Anal. Ed., 15, 325–6.

EXPERIMENT 2.   MOLECULAR DISTILLATION (3 periods)

OBJECTIVES

This experiment will provide experience with operations at high vacuum and the use of a molecular still.

PRINCIPAL EQUIPMENT AND SUPPLIES

Molecular still
Oil diffusion pump
Mechanical vacuum pump
Thermocouple gages for pressure measurement in the range of 0.05–100μ
Pure glycerol mono- and tri-palmitates

PROCEDURE

Much of the valuable experience that may be gained from this experiment lies in devising solutions to problems that arise in the installation of the vacuum system. Proper installation of traps, balance of forepump and oil diffusion pump capacities, choice of thermocouple gages, creation of a vacuum-tight system, and other problems will be encountered. For these reasons, the student will assemble the vacuum system as well as perform a molecular distillation and analyze the products.

For purposes of the experiment, you may separate a mixture of glycerol mono- and tri-palmitates, or you are at liberty to suggest an alternative mixture for distillation. Consult the appropriate industrial brochure for installation and operation of the molecular still available for your experiment. Consult ref. (1) for analytical procedures if glycerides are distilled.

TREATMENT OF DATA

Summarize your results and deliver a formal, oral report to the class.

Reference

1. Martin, J. B. 1953. The Equilibrium between Symmetrical and Unsymmetrical Monoglycerides and Determination of Total Monoglycerides. J. Am. Chem. Soc., 75, 5483–6.

B. Countercurrent Distribution

Countercurrent distribution is a relatively recent technique (1) for the fractionation of complex mixtures and the isolation of compounds such as antibiotics, hormones, etc., that might be damaged by the extremes of temperature or pH that occur in some other separation processes. Separations may be performed on amounts which range from the minimum detectable quantity up to as much as several hundred milligrams for certain kinds of compounds. In addition to the separation of mixtures, this process may serve as a criterion of identity and homogeneity of a preparation. The distribution constant of a compound between two given solvents is as reliable a characteristic as other physical properties (melting point, refractive index, absorption spectrum, etc.), whereas any departure from symmetry (skew) of the distribution curve is a fairly sensitive indication of inhomogeneity. Both theoretical and technical aspects of countercurrent distribution are covered in reviews (2, 3).

The separation of compounds by the countercurrent distribution machine is a cascade-type distribution between two liquid phases. The distribution is expressed by the partition factor (K') which is defined as the ratio of the total amount of solute in the moving phase to the total amount in the stationary phase. The partition factor in turn is related to the relative volumes of the two phases and to the more fundamental distribution constant (K) which is defined as the ratio of concentration of the solute in the moving phase to that in the stationary phase.

$$K' = \frac{c_1}{c_2} \cdot \frac{v_1}{v_2} = K \cdot \frac{v_1}{v_2} = K \cdot \alpha$$

where c_1 and c_2 are the concentrations in the two phases at equilibrium, v_1 and v_2 are the volumes of the two phases, and α is the ratio of v_1/v_2. The simplest case is the distribution of a single compound between equal volumes of the two liquid phases, i.e., $\alpha = 1.0$ and $K' = K$. For purposes of illustration let us further assume that at equilibrium the compound will distribute equally between the two phases. If we now equilibrate a second volume of phase 1 with the first volume of phase 2 and similarly equilibrate a second volume of phase 2 with the first volume of phase 1 we have the beginning of a systematic, cascade-type distribution. This type of liquid-liquid distribution is most efficiently accomplished in a countercurrent machine. The idealized results from four exchanges are shown in Table IV. For reasons that will become apparent below, the first tube is numbered zero.

TABLE IV. SIMPLIFIED DISTRIBUTION SCHEME

Transfer No.	Tube No.				
	0	1	2	3	4
0	.50/.50				
1	.25/.25	.25/.25			
2	.125/.125	.25/.25	.125/.125		
3	.0625/.0625	.187/.187	.187/.187	.0625/.0625	
4	.0312/.0312	.125/.125	.187/.187	.125/.125	.0312/.0312
Total contents of each tube after 4 exchanges	.062	.250	.375	.250	.062

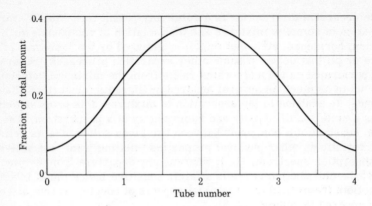

Figure 3. Normal distribution curve.

A curve of the type illustrated in Figure 3 is obtained when the tube number is plotted against the amount of compound in each tube. This curve is an example of a normal distribution curve, and it is expressed by an expansion of the binomial equation

$$(p + q)^n = 1.0 \qquad [1]$$

In this equation p is the amount of solute in the upper phase, q is the amount of solute in the lower phase, and n is the number of transfers or exchanges. For the case where the volumes of upper and lower phases are equal, the binomial equation [1] may be expressed in terms of K′ by the following transformations:

$$K' = \frac{p}{1 - p} \qquad \text{and similarly} \qquad K' = \frac{1 - q}{q}$$

$$K' - K'p = p \qquad\qquad K'q + q = 1$$

$$p = \frac{K'}{1 + K'} \qquad\qquad q = \frac{1}{1 + K'}$$

and by substitution in equation [1] for p and q

$$\left(\frac{K'}{1 + K'} + \frac{1}{1 + K'}\right)^n = 1.0 \qquad [2]$$

Expansion of binomial equation [2] for any term, $T_{n,r}$ (the total amount of material in the rth tube in a system of n transfers) gives

$$T_{n,r} = \frac{n!}{r!(n - r)!}\left(\frac{1}{K' + 1}\right)^n K^r \qquad [3]$$

The binomial expansion in the form of equation [3] could be used to plot the theoretical curve for the data in Figure 3. However, for the usual countercurrent distribution experiment where from twenty-four to several hundred transfers are performed, calculation of the theoretical curve by equation [3] becomes quite tedious. In order to simplify the calculations, one may assume that the experimental curve is a continuous function. Then it may be represented by equation [4] which is basically the general statistical formula for a normal curve of error (2).

$$y = \frac{1}{\sqrt{2\pi nK'/(K' + 1)^2}} e^{-x^2[2nK'/(K' + 1)^2]} \qquad [4]$$

In equation [4] y is the fraction of the substance in a given tube and x is the distance (in tube numbers) of this particular tube from the maximum n. Term n represents the position of maximum solute concentration. The value of n need not be an integral tube number, since it is calculated on the assumption that the distribution curve is a continuous function.

Values for K' and n must be determined from the original data in order that you may fit a normal distribution curve to your data by equation [4]. K' values may be calculated from the amount of solute in adjacent tubes by equation [5].

$$K' = \frac{r}{n + 1 - r}(T_r/T_{r-1}) \qquad [5]$$

A series of values for K' may be calculated by equation [5] with successive sets of data for the solute content of adjacent tubes $(T_r$ and $T_{r-1})$. The average K' value may be taken for calculation of a theoretical curve by equation [4], while any trend in the values of K' over the range of the curve may indicate inhomogeneity of the fraction. The average value of K' also may be employed to calculate the maximum, N, from the equation

$$N = \frac{nK'}{K' + 1} \qquad [6]$$

Study of equation [4] will reveal that its use for calculation of the theoretical curve is not so formidable as might appear at first. Certain terms in the equation are constant for a given set of data; while an initial calculation of the point y_0 (where r = o) will simplify the calculation of all other values of y.

Systematic departure of experimental data from the theoretical curve is observed as a distortion or skew of the normal distribution curve. A skew most frequently is caused by the presence of a second component. However, one should be on guard for other causes of skew; these include association or polymerization of the solute, change in distribution constant with change in temperature or change in concentration of the solute, complex formation between components of the original mixture, or operational errors such as the presence of emulsions during transfer, inequalities of volumes, etc.

References

1. Craig, L. C. 1944. Identification of Small Amounts of Organic Compounds by Distribution Studies. J. Biol. Chem., 155, 519–34.

2. Craig, L. C. 1950. Extraction and Distribution. In Technique of Organic Chemistry, III, 171–311. A. Weissberger, editor. Interscience Publishers. New York.
3. Weisiger, J. R. 1954. Countercurrent Distribution. In Organic Analysis, II, 277–326. Interscience Publishers, Inc. New York.

EXPERIMENT 3.   COUNTERCURRENT SEPARATION
OF PHTHALOYL-AMINO ACIDS (3 periods)

OBJECTIVES

There are three objectives in this experiment: (1) to become acquainted with the operation of the Craig countercurrent distribution machine; (2) to gain familiarity with calculations of the distribution constant and the normal distribution curve; and (3) to obtain experience in the recovery of small amounts of compounds.

PRINCIPAL EQUIPMENT AND SUPPLIES

Craig 25 tube countercurrent distribution machine (ref. [1], p. 265–71)*
Spectrophotometer
Silica spectrophotometer cells
pH meter
Apparatus for paper chromatography by the Redfield procedure (see Experiment 10)
"Unknown" mixture of phthaloyl-amino acids†
Ethyl acetate and n-butanol‡
1.0 M NaH_2PO_4
1.0 $\overline{M}$ K_2HPO_4
8 $\overline{M}$ H_3PO_4
Distilled HCl (20%)

PROCEDURE

The phthaloyl-amino acid mixture will be separated by a 24-transfer countercurrent distribution between two liquid phases. All fractions will be transferred to the upper phase for spectrophotometric analysis. Finally, the pure, crystalline phthaloyl-amino acids will be recovered and identified by melting point values.

1. *Countercurrent Distribution*

a. *Solvents.* An aqueous 1.0 $\underline{M}$ phosphate solution of pH 5.5 is prepared by mixture of 85 volumes of 1.0 $\overline{M}$ NaH_2PO_4 to 15 volumes of 1.0 $\underline{M}$ K_2HPO_4. Test the pH of this solution with the glass electrode pH meter and make

---

*A simpler apparatus which may be assembled from test tubes and metal rods is described in ref. (1), p. 259–60. See also ref. (2).

†Phthaloyl-amino acids are readily prepared (3). The simple monoamino, monocarboxylic acids distribute satisfactorily when ethyl acetate is the organic phase; the monoamino, dicarboxylic acids distribute satisfactorily when n-butanol is the organic phase. The dibasic, monocarboxylic acids do not distribute well in this system.

‡Be sure that the organic solvent does not contain impurities that absorb at 300 mμ. Different lots of the solvents vary, even when obtained from the same manufacturer; however, the impurities can be removed by careful fractional distillation.

necessary adjustments for pH 5.5. The organic solvent is ethyl acetate or n-butanol.

b. *Equilibration of phases.* Saturate 250 ml portions of the aqueous and organic solvents with one another by shaking them together for several minutes in a separatory funnel. Allow the mixture to stand until the phases separate; then draw off the lower phase. Henceforth, the upper (organic) phase will be termed equilibrated upper phase, and the lower (aqueous) phase will be termed equilibrated lower phase.

c. *Preparation of the phthaloyl-amino acid sample.* Dissolve 20 mg of the unknown phthaloyl-amino acid mixture in 5 ml of equilibrated upper phase.

d. *Loading the countercurrent distribution machine.* Liquid seals of this machine depend upon perfectly clean and unmarred surfaces between top plate, top cylinder, and bottom cylinder. Remember that the stainless steel and aluminum used in the construction of this machine are readily scored. Wipe the metal surfaces with the palm of the hand before assembling the machine. This will remove any lint or dust. Be sure that locks A, D, and E are tight (Figure 4). Turn top cylinder so that an upper tube spans the wall between two lower tubes and is thus in direct connection with two lower tubes. In this position, all lower tubes are in direct connection. Through the hole in the top glass plate add 205 ml of equilibrated lower phase to any tube. Rotate the top cylinder several times to distribute the lower phase evenly among all tubes.

Line up the top cylinder so that upper tube no. 0 is directly over lower tube no. 0. Now introduce 8 ml of equilibrated upper phase into each tube in the top cylinder, except tube no. 0. Introduce the phthaloyl-amino acid sample into tube no. 0. Rinse the vessel which had contained the sample with 3 ml of equilibrated upper phase, and add this also to tube no. 0. Tube no. 0 should now contain the phthaloyl-amino acid sample in 8 ml of equilibrated upper phase. Close the hole in the top plate after all solutions have been added.

e. *Countercurrent distribution:*

(1) Tighten lock B slowly but firmly to seal the machine. Care should be exercised at this point since excessive pressure may crack the glass plates.
(2) Loosen lock D on the spindle to release the machine.
(3) Proceed with the distribution by 50 complete inversions of the machine.
(4) Tighten lock D and allow time for phases to separate.
(5) Loosen lock B.
(6) Release lock C and turn top cylinder by one tube in a clockwise direction.
(7) Set lock C.

Now the upper phase of tube no. 0 is over the lower phase of tube no. 1, and a fresh upper phase is over the lower phase of tube no. 0

Repeat this distribution cycle until the original upper phase (top tube no. 0) is over the 24th lower phase (bottom tube no. 24). You have now completed a 24-transfer distribution.

f. *Removal of solutions.* Loosen lock B so that the top glass plate can be turned. Through the hole in this plate add 1 ml of $8\,M\,H_3PO_4$ to each tube. Tighten lock B and mix contents. The phthaloyl-amino acids will now pass completely into the upper phase. Withdraw both phases from

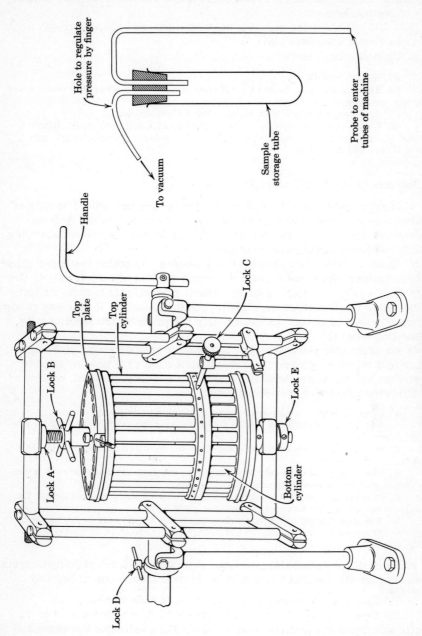

Figure 4. Countercurrent distribution machine.

each tube by means of a suction tube (Figure 4) and store in well-stoppered test tubes. Sample numbers should correspond with the <u>lower</u> set of numbers of the machine.

<u>Caution</u>: Do not allow the machine to remain exposed to the strongly acid solutions longer than necessary.

g. *Cleaning the machine:*

(1) Secure lock D.

(2) Release locks A, B, and C, and then remove the top bar, top plate, glass plate and top cylinder.

(3) Release lock E and then remove the bottom cylinder.

(4) Wash each part separately and carefully with mild soapsuds. Rinse thoroughly with water and finally with acetone. Allow the apparatus to dry completely before reassembling.

2. *Analysis of the Distribution Fractions*

a. *Determination of optical density.* Determine the optical density of an aliquot of the upper phase from each tube at 300 mμ in the Beckman spectrophotometer. Use silica cells, and balance the instrument against a "blank" of equilibrated organic phase.

b. *Identification of the distribution fractions.* Combine the upper phase of three or four tubes that contain the major portion of each fraction. Dry these fractions over $CaSO_4$ and filter them. Concentrate twenty- or fortyfold in order to crystallize the phthaloyl-amino acid (4). If necessary, add petroleum ether to reduce solubility. Dry the product, and identify it by the melting point (3).

In the event that you are unable to separate the particular phthaloyl-amino acid mixture by countercurrent distribution or if your attempt to crystallize the separate phthaloyl-amino acids is unsuccessful, employ the following procedure for identification.

1. Evaporate the phthaloyl-amino acid solution to dryness.

2. Add 1.0 ml of distilled 20% HCl and transfer the mixture to a small tube sealed at one end.

3. Seal the tube under vacuum; then heat the tube and its contents at 110° for 6 hr.

4. When cool, open the tube and evaporate the contents to dryness in a vacuum desiccator or vacuum oven over NaOH pellets.

5. Extract the amino acid with a small volume of water and centrifuge the mixture in the same tube to remove insoluble phthalic acid.

6. The supernatant solution which contains the amino acid is spotted on paper for identification by the paper chromatographic procedure of Redfield (see p. 52).

c. *Analysis of distribution curves.* Distribution of the phthaloyl-amino acids between the various tubes may be represented in one of two ways:

(1) Plot optical density at 300 mμ versus tube number.

(2) Determine the molar extinction coefficient at 300 mμ of an authentic sample of the phthaloyl-amino acid. Then calculate the amount of phthaloyl-amino acid in each tube of the distribution from the extinction coefficient, optical density, and solution volume. Finally, plot the amount of phthaloyl-amino acid versus tube number.

In all cases where the separation of fractions will permit, determine the distribution coefficients of the phthaloyl-amino acids (K'), the maxima (N), and plot the theoretical normal distribution curve from these values of K' and N.

TREATMENT OF DATA

Prepare a brief, written report that includes all of the results in tabular and graphic form, as well as the identity and relative amounts of components of the "unknown," pertinent comments, deviations from prescribed procedure, etc.

In the event that a number of phthaloyl-amino acids were used in preparation of the samples, the class may compare results and discuss the distribution behavior of the various phthaloyl-amino acids.

QUESTIONS

1. A good distribution of phthaloyl-histidine cannot be obtained under the conditions described in this experiment. Why should this be so?

2. Pure reference amino acids and amino acids from the hydrolysis of the separated phthaloyl-amino acids in this experiment frequently do not travel at the same rate on a paper chromatogram. Why is there a difference in rate, even if the amino acids are the same in both samples? How could you make a better comparison by paper chromatography which is not subject to this interference?

3. The distribution in this experiment was conducted at pH 5.5; what effect would you predict a change to pH 3 would have on the distribution?

4. Why is the first tube of the countercurrent distribution machine numbered zero?

References

1. Craig, L. C. 1950. Extraction and Distribution. In Technique of Organic Chemistry, III, 171–311. A. Weissberger, editor. Interscience Publishers. New York.
2. Pinsky, A. 1955. Simple Countercurrent Distribution Apparatus. Anal. Chem., 27, 2019–20.
3. Billman, J. H., and Harting, W. F. 1948. Amino Acids. V. Phthalyl Derivatives. J. Am. Chem. Soc., 70, 1473–4.
4. McElvain, S. M. 1946. The Characterization of Organic Compounds, 6–12. Macmillan Co. New York.

EXPERIMENT 4.   COUNTERCURRENT SEPARATION OF METHYL
ESTERS OF FATTY ACIDS (4 – 5 periods)

OBJECTIVE

This experiment will expose the student to a more difficult separation than was encountered in Experiment 3. The experiment also will permit comparison between distillation and countercurrent distribution operations for the separation of methyl esters of fatty acids.

PRINCIPAL EQUIPMENT AND SUPPLIES

All-glass, 200-tube, countercurrent distribution machine in a room at constant temperature
Mixture of methyl esters of fatty acids
Hydrocarbons (pentane-hexane)
Nitroparaffins (nitromethane-nitroethane)

PROCEDURE

A good separation by countercurrent distribution of a mixture of esters of the fatty acids will require many more distributions than were made in Experiment 3. A more complicated distribution machine must be used. Further, the solvent system employed is highly sensitive to temperature changes. These and other complications exist that were not encountered in the preceding experiment.

Consult the article of Cannon et al. (1) for details.

TREATMENT OF DATA

Summarize your results and deliver a formal, oral report to the class.

Reference

1. Cannon, J. A., Zilch, K. T., and Dutton, H. J. 1952. Countercurrent Distribution of Methyl Esters of Higher Fat Acids. Anal. Chem., 24, 1530–2.

C. Column Chromatography

The first outstanding application of column chromatography to scientific research was the separation of plant pigments on columns of powdered calcium carbonate by Tswett in 1910.* Tswett foresaw wider applications for this process of adsorption chromatography and even envisioned its use for separations of colorless materials. Indeed, the simplicity of apparatus, the efficiency of separation, and the mild conditions have led to an extension of column chromatography to separations based on processes other than that of adsorption. These now include gas-liquid partition chromatography (1), liquid-liquid partition chromatography (2), and ion exchange chromatography (3, 4).

Each of these techniques involves a continuous repetition of a fundamental separation process (see Table I, p. 6), and all processes have in common a transfer of solute from one phase to another. Adsorption chromatography is a continuous process of adsorption-desorption of solutes to and from a gas or liquid phase and the surface of the solid packing material of the column. The rate of movement of solute down the column may be treated theoretically by equations derived for batchwise adsorption (5). Ion exchange chromatography is a continuous, competitive exchange of ions in solution for those electrostatically bound to oppositely charged groups on the resin, and the rate of movement of a given ionizable compound down the column is a function of its degree of ionization, the concentrations of other ions, and the relative affinities of the various ions present in the solution for charged sites on the resin (4). Partition chromatography ideally represents a continuous process of distribution and redistribution of solute between two liquid phases, or a liquid and a gas phase, as the solute moves down the column. One phase is mobile and serves to carry the solute down the column; the second, liquid phase is immobile for it is held in the interstices of the solid-packing material of the column. The second phase tends to retain the solute and thus to retard the rate of migration of the solute down the column. Hence, the rate of movement of a solute during partition chromatography is a function of the distribution constant of the solute between the two phases (1, 2).

Amounts of material from one mg or less up to as much as a gram or more may be handled in laboratory scale columns and one or another of these column chromatographic techniques has been used for separations involving almost every kind of biological compound (Table V) as well as for analysis of mixtures and identification of components of mixtures. For the analysis of mixtures, the manner of operation of the column is an important factor and governs in large part the kind of results that are obtained. For example, any one of the various techniques of frontal analysis, displacement analysis, elution analysis, and gradient elution analysis (5, 6) may be employed for the analysis of mixtures by adsorption chromatog-

---

* Techniques of column chromatography probably had been employed for centuries by artisans; witness the description of an ion exchange desalting of sea water in a 17th century novel by Daniel Defoe entitled The Life, Adventures, and Piracies of Captain Singleton.

TABLE V. REPRESENTATIVE CLASSES OF COMPOUNDS THAT
HAVE BEEN SEPARATED BY COLUMN CHROMATOGRAPHY

Adsorption Chromatography	Partition Chromatography	Ion Exchange Chromatography
Carbohydrates	Amines	Amino acids
Carotenoids	Esters	Coenzymes
Porphyrins	Fatty acids	Nucleotides
Sterols	Hydrocarbons	Peptides
	Proteins	

raphy. The choice of separation process, nature of the elution mixture, type of packing, and method for analysis of products rests with the physical and chemical properties of the components of the given mixture. Proper choice of these conditions is the most difficult aspect of the problem and often the choice must be an empirical one.

References

1. James, A. T., and Martin, A. J. P. 1952. Gas–Liquid Partition Chromatography: The Separation and Micro-estimation of Volatile Fatty Acids from Formic Acid to Dodecanoic Acid. Biochem. J., 50, 679–90.
2. Martin, A. J. P., and Synge, R. L. M. 1941. A New Form of Chromatogram Employing Two Liquid Phases. Biochem. J., 35, 1358–68.
3. a. Kunin, R. 1950. Ion Exchange Resins. John Wiley and Sons. New York.
   b. Osborn, G. H. 1956. Synthetic Ion-Exchangers. Macmillan Co. New York.
4. Nachod, F. C., and Schubert, J., editors. 1956. Ion Exchange Technology. Academic Press. New York.
5. Cassidy, H. G. 1951. Adsorption and Chromatography. In Technique of Organic Chemistry, V, 213–227. A. Weissberger, editor. Interscience Publishers. New York.
6. Williams, R. J. P. 1954. General Principles of Chromatography. Brit. Med. Bull., 10, 165–9.

EXPERIMENT 5.   ION EXCHANGE CHROMATOGRAPHY APPLIED TO THE SEPARATION OF NUCLEOTIDES (3 periods)

OBJECTIVE

This experiment will provide experience in the use of ion exchange resins for the separation of ionizable compounds.

PRINCIPAL EQUIPMENT AND SUPPLIES

Fraction collector
Spectrophotometer
Silica spectrophotometer cells
Sample bottles (100/student); 8-dram, 24 by 95 mm, screwcap vials serve well
"Unknown" mixture of nucleotides that may also contain purines, pyrimidines, and/or nucleosides
Dowex 1 (Cl$^-$) (200–400 mesh; 8X) washed and free of finer particles
1 M formic acid
1 $\overline{\text{M}}$ sodium formate
0.$\overline{2}$ N HCl

PROCEDURE

Ion exchange resins may be employed to separate purine and pyrimidine bases and the nucleosides as well as nucleotides (1). The present procedure, however, will separate only the nucleotides from one another and from any purine and pyrimidine bases and nucleosides present. As a consequence, your analyses of the unknown mixture will yield values for the amount of each individual nucleotide and an estimate of the total content of purine and pyrimidine bases plus the nucleosides.

1. *Preparation of the Ion Exchange Column*

a. *Formation of the ion exchange column.* Make a glass column as shown in Figure 5 (p. 28). The inside diameter of the tube should be 1.4 cm and the main body of the tube should be 20–25 cm long. Place a small plug of glass wool in the constriction of the tube to retain the resin. Shake the stock suspension of Dowex 1 (Cl$^-$) and add the slurry by pipet to the tube. Allow the column to drain and the resin to settle by gravity, but do not let the liquid level drop below the level of the resin. (Note: This latter precaution must be observed throughout the use of the column. If at any time the liquid falls below the top of the resin phase, air may enter and be entrapped. These air bubbles result in channeling and render the column useless. If the liquid level does fall below the resin surface during the preparation of the column, add more liquid and stir the top resin layers to remove air bubbles. This measure cannot be applied, however, once the column is in use for ion separation.) Continue to add the resin slurry until the column of resin is 10 cm high (15.4 cm^3 of resin bed).

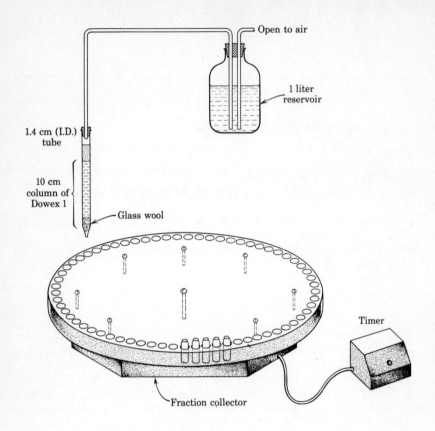

Open to air

1 liter
reservoir

1.4 cm (I.D.)
tube

10 cm
column of
Dowex 1

Glass wool

Timer

Fraction collector

Figure 5. Ion exchange column and fraction collector.

b. *Conversion of the resin to the formate form by the following series of washes.*

(1) Pass 1 M sodium formate slowly down the column (approx. 1 ml/min) until the effluent is free of chloride ion (AgNO$_3$ test).

(2) Pass approximately 50 ml of 1 M formic acid down the column.

(3) Wash thoroughly with distilled water until the effluent is at the pH of distilled water.

By this series of washes the following changes have occurred on and in the resin (R):

$$R^+Cl^- + Na^+ + HCOO^- \text{ (excess)} \longrightarrow R^+HCOO^- + \underline{Na^+ + Cl^- + HCOO^-}$$

Removed in the
various washes

2. *Introduction of the Nucleotide Mixture onto the Column*

The solution of the nucleotide "unknown" will contain a total of from 6 to 10 mg of bases, nucleosides, and nucleotides in a volume of about 1 ml. (This represents approximately 10% of the total nucleotide capacity of the 15 cm^3 of resin.) Adjust the nucleotide mixture to pH 8 or higher.

Mount the liquid reservoir and the resin column above the fraction collector. Load the collector with 50−60 clean, dry bottles of 30 ml capacity. Allow the liquid level of the column to drop within approximately 1 mm of the resin level; then add the nucleotide mixture cautiously by pipet.

3. *Elution of the Nucleotide Mixture*

Allow liquid to flow at an effluent rate of 1 ml/min. When the liquid level is again 1 mm above the resin level, rinse the sample bottle and sides of the tube with 2−4 ml of water. Continue the effluent flow until the liquid is 1 mm above the resin; then add water to increase the level to 5−6 cm above the resin. Place 50 ml of water in the reservoir and continue elution at a flow rate of 1 ml/min. The fraction collector will be set to change samples every 25 min; therefore, each fraction will have a volume of 25 ml.

When almost all of the water has passed from the reservoir, replenish the reservoir in turn with 500 ml of 0.1 M formic acid (pH 2.3−2.4); 500 ml of 0.25 M formic acid; 500 ml of 1.0 M formic acid (pH 1.8); and finally 250 ml of 0.2 N HCl. This procedure is a rather inefficient one and is rendered necessary by heavy demand for the spectrophotometer. In normal practice, one would follow the eluent composition closely and make changes in the solvent in accordance with the appearance of nucleotide fractions in the eluent. The solvent flow may be stopped without deleterious effects at any time when the column must be left unattended for a considerable interval of time, such as overnight.

4. *Analysis of Fractions from the Column*

Measure the optical density of each fraction in the spectrophotometer at 260 mμ. Use formic acid or hydrochloric acid of corresponding concentrations as the blank for adjustment of the instrument. A plot of optical density as ordinate and volume of eluent (or tube number) as abcissa will yield a "profile" of the fractionation.

Identify the nucleotide in each peak on the basis of characteristic ratios of optical densities at various wave lengths and pH values (see Table VI and ref. [2]), and the relative rates of elution (see Figure 6, p. 31). Identification of the isomeric 2', 3', and 5' nucleotides would require paper chromatography (3) or specific enzymatic tests (4).

Calculate the amount of each component from the molar extinction coefficients (Table VI), the observed optical densities, and the volumes of the fractions. For an estimate of the amount of bases and nucleosides that may have appeared in the early fractions, assume a molar absorption coefficient at 260 mμ of 10,000.

TABLE VI. CHARACTERISTICS OF ABSORPTION SPECTRA OF SOME PURINES, PYRIMIDINES, NUCLEOSIDES, AND NUCLEOTIDES(2)

Compound	Mol. Weight	Spectra in Acid*				Spectra in Alkali†			
		λ_{max}	$E_{max} \times 10^3$	$\frac{250}{260}$	$\frac{280}{260}$	λ_{max}	$E_{max} \times 10^3$	$\frac{250}{260}$	$\frac{280}{260}$
Purines									
Adenine	135.11	263	13.2	0.76	C.38	269	12.3	0.57	0.60
Guanine	151.15	276	7.4	1.37	C.84	274	8.0	0.99	1.14
Xanthine	152.11	267	10.3	0.57	0.61	278	9.3	1.29	1.71
Pyrimidines									
Cytosine	111.10	276	10.0	0.48	1.53	282	7.9	0.60	3.28
Thymine	126.11	265	7.9	0.67	C.53	291	5.4	0.65	1.31
Nucleosides									
Adenosine	267.24	257	14.6	0.84	0.22	260	14.9	0.78	0.14
Guanosine	283.26	256	12.2	0.94	0.70	258–266	11.3	0.89	0.61
Cytidine	243.23	280	13.4	0.45	2.10	271	9.1	0.87	0.93
Inosine	268.24	249	12.2	1.68	0.25	253	13.1	1.05	0.18
Nucleotides									
Adenylic acid (2')(a)	347.22	257	14.4	0.85	0.23	259	15.4	0.80	0.15
Adenylic acid (3')(b)	347.22	257	14.4	0.85	0.22	259	15.4	0.80	0.15
Adenylic acid (5')	347.22	257	15.1	0.84	0.22	259	15.4	0.79	0.15
Guanylic acid (5')	363.24	257	12.2	0.99	0.70	256	11.1	0.89	0.60
Cytidylic acid (5')	323.21	281	13.6	0.46	2.10	274	7.3	0.84	0.99
Inosinic acid (5')	348.22	(247)				(254)	(13.5)		
Uridylic acid (5')	324.18	261	9.7	0.74	0.38	261	7.30	0.82	0.33

* pH 1–3
† pH 11–13

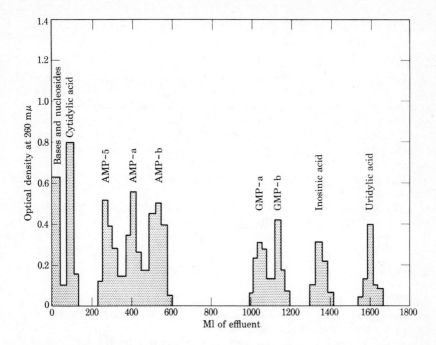

Figure 6. Schematic representation of the probable order of elution of nucleotides by the procedure in Experiment 5.

5. *Recovery of a Nucleotide from the Eluent* (Optional)

One advantage of the use of formic acid solutions for elution is that simple evaporation may be used to recover the products. Pool the fractions containing the nucleotide that you wish to isolate, and concentrate them in vacuo to 5 – 10 ml. Transfer the concentrate quantitatively into a tared, 10 – 25 ml tube and concentrate to dryness in vacuo. Reweigh the tube. It would be well to repeat this procedure with a similar volume of eluent that contains no nucleotide in order to determine non-volatile materials other than the nucleotide in the eluent. Compare the observed weight of nucleotide with the value calculated from optical densities.

TREATMENT OF DATA

Assume that your nucleotide mixture was the hydrolysis product from a nucleic acid fragment isolated from yeast cells, and assume that you wish to publish the results of this experiment in the form of a Note to appear in the Journal of the American Chemical Society. On the basis of these assumptions, prepare a manuscript in the form required by the editors of this journal with the exception that only a single copy need be made.

Consult the front matter of J. Am. Chem. Soc., Vol. 78, No. 3, for particulars. The detailed procedures that appear in this Experiment need not be reproduced but may be cited by reference.

QUESTIONS

1. Two separate properties of the eluate were varied during elution of the nucleotides by the recommended procedure. What was the effect of each of these changes on the rate of elution of the nucleotides?

2. The recommended resin was Dowex 1 (formate) (200–400 mesh; 8X). What does this currently accepted shorthand tell you of the properties of the resin? Why was a resin of these properties chosen for this separation?

3. How might you separate purine and pyrimidine bases on an ion exchange column?

4. Why was the Dowex 1 (Cl$^-$) converted to the formate form with a solution of sodium formate rather than formic acid?

References

1. Cohn, W. E. 1955. Separation of Nucleic Acid Derivatives by Chromatography on Ion-Exchange Columns. In The Nucleic Acids, I, 211–41. E. Chargaff and J. N. Davidson, editors. Academic Press. New York.
2. Beaven, G. H., Holiday, E. R., and Johnson, E. A. 1955. Optical Properties of Nucleic Acids and Their Components. In The Nucleic Acids, I, 493–553. E. Chargaff and J. N. Davidson, editors. Academic Press. New York.
3. Wyatt, G. R. 1955. Separation of Nucleic Acid Components by Chromatography on Filter Paper. In The Nucleic Acids, I, 243–65. E. Chargaff and J. N. Davidson, editors. Academic Press. New York.
4. Schmidt, G. 1955. Nucleases and Enzymes Attacking Nucleic Acid Components. In The Nucleic Acids, I, 555–625. E. Chargaff and J. N. Davidson, editors. Academic Press. New York.

EXPERIMENT 6. LIQUID-LIQUID PARTITION CHROMATOGRAPHY APPLIED TO THE SEPARATION OF THE LOWER FATTY ACIDS (2 periods)

OBJECTIVE

This experiment will serve as an introduction to the general technique of column, liquid-liquid partition chromatography.

PRINCIPAL EQUIPMENT AND SUPPLIES (1)

Chromatography tube (18 mm by 40 cm)
Steam distillation apparatus
Mixture of fatty acids $(C_2 - C_5)$
Silica gel
Purified butanol and chloroform
Bromcresol green

PROCEDURE

Partition chromatography has proved to be a valuable technique for the separation of mixtures of small amounts (less than one gram) of volatile fatty acids. Both the techniques of gas-liquid partition and liquid-liquid partitition have been used with such mixtures. In the present instance, an aqueous or sulfuric acid phase will be held on a column of silica gel or Celite. The fatty acid mixture will be introduced and swept down this column by a second, organic phase. Fractionation of the mixture will depend upon differences in the distribution coefficients of the fatty acids. You may follow either the procedure of Elsden (1) or that of Peterson and Johnson (2). If the procedure of Elsden is chosen, you may follow the course of the fatty acids down the column by the change in color of the bromcresol green on the column. Identification of the bands may be made by comparison of their R values with R values for known compounds.

$$R = \frac{\text{cm moved by the band}}{\text{cm moved by the surface of the developing solvent in the tube above the chromatogram}}$$

Representative values obtained under the conditions used by Elsden (1) are given in Table VII (Note: these values change slightly with change in amount of acids in the sample).

TABLE VII. R VALUES FOR LOWER FATTY ACIDS (1)

Fatty Acid (0.02 millimole)	R
Acetic	0.32
Propionic	0.76
n-Butyric	1.09
n-Valeric	1.33

TREATMENT OF DATA

Summarize your results and deliver a formal, oral report to the class.

References

1. Elsden, S. R. 1946. The Application of the Silica Gel Partition Chromatogram to the Estimation of Volatile Fatty Acids. Biochem. J., 40, 252–6. See also Biochem. Soc. Symposium No. 3, p. 74 (1949).
2. Peterson, M. H., and Johnson, M. J. 1948. The Separation of Fatty Acids of Intermediate Chain Length by Partition Chromatography. J. Biol. Chem., 174, 775–89.

EXPERIMENT 7. GAS-LIQUID PARTITION CHROMATOGRAPHY APPLIED TO THE SEPARATION OF VOLATILE FATTY ACIDS (2 periods)

OBJECTIVE

You should acquire from this experiment some experience in the assembly and operation of apparatus for gas-liquid partition chromatography.

PRINCIPAL EQUIPMENT AND SUPPLIES

Glass column and ancillary equipment for the partition column (see Figure 7; this equipment may be readily constructed from directions in ref. [1]).
Cylinder of nitrogen gas
Mixture of fatty acids
Celite; acid-washed and graded
Silicone; Dow-Corning 550
Stearic acid
Ethyl Cellosolve
0.005 N NaOH

PROCEDURE

Details for assembly and operation of the gas-liquid partition column shown in Figure 7 are admirably set forth in ref. (1). A separation of

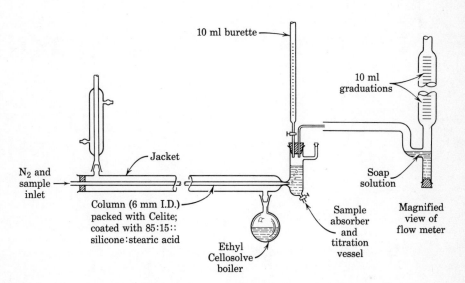

Figure 7. Apparatus for gas-liquid partition chromatography.

volatile fatty acids is suggested; however, a separation of amines (2), methyl esters of the fatty acids (3), or other volatile compounds may be attempted.

TREATMENT OF DATA

Summarize your results and deliver a formal, oral report to the class.

References

1. James, A. T., and Martin, A. J. P. 1952. Gas-Liquid Partition Chromatography: the Separation and Micro-estimation of Volatile Fatty Acids from Formic Acid to Dodecanoic Acid. Biochem. J., 50, 679–90.
2. James, A. T., Martin, A. J. P., and Smith, G. H. 1952. Gas-Liquid Partition Chromatography: the Separation and Micro-estimation of Ammonia and the Methyl-amines. Biochem. J., 52, 238–47.
3. James, A. T., and Martin, A. J. P. 1956. Gas-Liquid Chromatography: the Separation and Identification of the Methyl Esters of Saturated and Unsaturated Acids from Formic to n-Octadecanoic Acid. Biochem. J., 63, 144–52.

EXPERIMENT 8. ADSORPTION CHROMATOGRAPHY APPLIED TO THE SEPARATION OF CAROTENES (3 periods)

OBJECTIVE

This experiment will illustrate the application of adsorption chromatography to the separation of labile compounds.

PRINCIPAL EQUIPMENT AND SUPPLIES

1. Separation of Crude Carotenes from Other Plant Pigments
   Fresh spinach (28 oz)
   Petroleum ether (Skelly Solve "B") (B.P. 60 – 70)
   Calcium hydride
   U.S.P. calcium carbonate or confectioner's sugar
   Chromatography tube (3 by 25 cm)

2. Separation of Carotene Isomers
   Melting point apparatus (optional)
   Polarimeter (optional)
   Spectrophotometer (optional)
   Crude carotene fraction from Part 1 (25 mg) or commercial carotene (mixed isomers)
   Magnesium oxide (Merck, U.S.P., Heavy)
   Calcium hydroxide or magnesium oxide (Micron brand)
   Hyflo Super Cel
   Freshly redistilled petroleum ether
   Chromatography tube (2 by 45 cm)

PROCEDURE

1. *Separation of Crude Carotenes from Other Pigments*

The experiment described is very similar to one performed by Tswett in 1905 and it can be carried out in the following way (1). A suitable quantity (28 oz) of fresh spinach* is ground in a large mortar with 200 ml of 95% alcohol and about twice as much petroleum ether. Acetone may be used instead of alcohol, but one or the other polar solvent is necessary to help extract the pigments. The green extract is filtered through cheesecloth to separate the extract from the pulp. The extract is washed repeatedly with one-half its volume of water in a separatory funnel until the water layer remains clear. The funnel should be swirled gently instead of shaken, to avoid emulsion formation. It may be necessary to add petroleum ether (up to 300 ml) in order to form two phases if too much petroleum ether has evaporated during the preceding operations. The organic layer is now carefully separated from the aqueous layer, and the petroleum ether solution is evaporated to dryness with <u>mild</u> heat. (<u>Caution</u>: remove flames

---

* A mixture of equal amounts of spinach and carrots will increase the carotene fraction; the carrots should be ground in a blender before extraction.

or sparking electrical equipment.) If all the materials are on hand, this whole operation, from leaves to dried green pigment, should take approximately an hour. Store the dried pigments under vacuum in the dark since the carotenes are relatively unstable.

While the pigment extract is drying, a column of adsorbent is prepared. A chromatography tube of about 3 by 25 cm dimensions is fitted with a disk of filter paper over the perforated bottom plate. About $3-4$ gm of U.S.P. calcium carbonate or confectioner's sugar is put into the tube and very gently pressed down; it is important that the adsorbent be added in small portions and that it be packed gently but uniformly to the edges of the tube in order to prevent channels. A column of adsorbent about 15 cm in length is prepared in this manner.* The preparation of a chromatography column is an art in which proficiency is attained only by experience. Some common errors include: packing the adsorbent too tightly so that the solution passes through very slowly; packing the center more tightly than the sides so that the bands travel faster near the edge of the tube; generally uneven packing which results in jagged bands; and an irregular surface layer of adsorbent which starts the bands unevenly on their way down the column.

The leaf extract is dissolved in 15 ml of anhydrous petroleum ether (dried over CaH_2 or Na_2SO_4) just prior to chromatography, and CaH_2 or Na_2SO_4 is added to insure complete dryness. All polar solvent from the leaf extraction must be removed for successful chromatography. Approximately 20 to 25 ml of dried petroleum ether should first be introduced to the top of the column and allowed to run into the packing. The pigment extract is immediately added in one portion so that a sharp initial green zone is formed. If the column is packed too tightly, it may be necessary to apply air pressure to the top of the column. If the petroleum ether extract was not washed sufficiently or was not dried, the zone does not form and the green solution runs on through the tube. (The petroleum ether front should be considerably ahead of the green zone.) The success of the chromatogram often depends upon the dryness of the petroleum ether, the manner of packing the column, and the method of initial application of the pigment to the column.

When practically all the pigment solution has entered the column but before air has entered, several 10 ml washings of pure petroleum ether are applied and allowed to run down in a similar manner. This is then followed by the developer solvent. If the adsorbent is calcium carbonate, the developer is 50% benzene in petroleum ether; if the adsorbent is sugar, the developer is 25% benzene in petroleum ether. As the developer passes into the adsorbent, the green zone spreads and becomes differentiated into an upper thin, dark-green zone of chlorophyll b; a lower blue-green zone of chlorophyll a; an orange zone of the xanthophylls below this; and a fast-running, yellow zone of the carotenes still lower. The zone of carotenes

---

*If the column, packed in this manner, fails to give a satisfactory separation of the pigments an alternative packing may be tried. Suspend confectioner's sugar in anhydrous petroleum ether and add this to the empty column. Stir the suspension with a glass rod to remove any air bubbles and allow the sugar to settle by gravity. Drain the petroleum ether from the column and add the leaf extract.

may easily be washed out of the column by elution with the developer solvent. The other zones can be recovered by extruding the column, cutting it between the zones, and eluting each zone in a separate test tube with petroleum ether or with benzene containing a drop of alcohol. An alternate procedure for isolation of the carotenes is to develop the chromatogram until the carotene zone is well separated from the other zones. Nitrogen under pressure is then passed through the column to dry it, and the whole column is extruded or the bottom zone containing the carotenes may be dug out with a spatula. The carotene pigments are then extracted with a small amount of petroleum ether.

The eluent or extract that contains the carotenes is rapidly concentrated to dryness in vacuo and weighed. It is important that all operations be carried out in subdued light, and steps that involve carotene in solution should be done as rapidly as possible. Powdered or crystalline carotene may be stored for some time in the dark; for permanent storage it should be sealed in an evacuated tube and placed in the dark.

2. *Separation of Carotene Isomers*

A chromatography column is prepared by filling a 2 by 45 cm tube to a height of 30 cm with a 3 : 1 mixture of magnesium oxide (Merck, U.S.P., heavy) and Hyflo Super Cel filter aid. The tube is attached to a 500 ml suction flask, about 2 cm of glass wool is placed in the bottom, and the adsorbent is added in small portions under suction. Each portion of adsorbent is pressed firmly into place with a plunger which fits the bore of the tube rather closely. About twenty separate additions are made in all. Particular care should be used in packing the topmost layer of the filling, so as to have a uniform flat surface of even tightness. Finally, place a disk of filter paper on top of the column of adsorbent.

About 25 mg of the carotene fraction obtained above or 25 mg of commercial carotene (mixed isomers) is dissolved in the smallest possible amount of dry, freshly redistilled petroleum ether (about 30 − 35 ml). A small amount of pure petroleum ether is carefully poured into the adsorption tube and it is permitted to flow down into the packing until the solvent level is just above the top of the adsorbent. (Pressure may be necessary to obtain a uniform flow through the column.) The carotene solution is added, and this, in turn, is allowed to enter the column of adsorbent. Then several successive 5 ml portions of petroleum ether are added, so as to cause the entire sample to enter the top of the column in as narrow a band as possible.

·The tube is then filled with petroleum ether for development of the chromatogram. The chromatogram is developed until the three bands of α, β, and γ carotene are sufficiently far apart to permit effective separation. When this has been accomplished, the solvent level is allowed to drop below the top of the adsorbent, so that air is admitted to the column. The adsorbent will be seen immediately to shrink away from the walls of the tube, and as soon as this has happened (5 − 10 seconds after air is admitted), the suction is interrupted, the tube is inverted, and the adsorbent is slid out as a compact cylinder by gentle tapping or by pushing from the bottom end with a glass rod.

The beta-carotene zone (middle, orange-colored) is separated by carving the cylinder of adsorbent with a spatula, and is eluted with 5% ethanol in petroleum ether. The filtered eluate is concentrated to dryness in vacuum; the residue is dissolved in the minimum amount of warm benzene and is filtered if necessary. Two or three volumes of boiling methanol are added, and the mixture is kept in the cold room overnight. The crystals which form are filtered off with suction, washed with a little cold methanol, and dried over P_2O_5 at $60-80°$ and a pressure of 1 mm of Hg or less.

The product should melt at $183.5°$ (uncorrected, open tube), exhibit no optical activity in benzene, and have an absorption coefficient ($E_{1\ cm}^{1\%}$) of 2350 at 465 mμ in benzene* (2).

TREATMENT OF DATA

Summarize your data and deliver a formal, oral report to the class.

QUESTIONS

1. One criterion of purity of α, β, and γ carotenes is the absorption spectrum. Compare the spectra of these three isomers and evaluate the sensitivity of spectral studies for the detection of small amounts of α or γ carotene in a sample of β carotene.

2. Why are the columns used in this experiment so sensitive to traces of water?

References

1. Cassidy, H. G. 1951. Adsorption Chromatography. In Technique of Organic Chemistry, V, 320–322. A. Weissberger, editor. Interscience Publishers. New York.
2. Devine, J., Hunter, R. F., and Williams, N. E. 1945. The Preparation of β-Carotene of a High Degree of Purity. Biochem. J., 39, 5–6.

---

*Commercial, thiophene-free benzene, fractionated, and collected over a range of 0.2°.

EXPERIMENT 9. ION EXCHANGE RESINS AS ACID-BASE CATALYSTS (3 periods)

OBJECTIVE

This experiment will provide an introduction to the use of charged resins for reactions other than simple ion exchange.

PRINCIPAL EQUIPMENT AND SUPPLIES

Dowex 50 (H^+) as the dry powder
Ethyl acetate

PROCEDURE

Ion exchange resins, by virtue of their strongly acidic or basic groups, can catalyze reactions that are usually catalyzed by strong acids or bases in solution. These resins have the distinct advantage that the catalyst may be readily separated from the reaction mixture at any desired time. One example of the application of resins for study of biological materials is the employment of Dowex 50 (H^+) for the partial hydrolysis of adenylic acids by Khym et al. (1) in the course of their proof of the structures of adenylic acid isomers a and b. Other published work on this property of resins is scanty (2–6). The student may use Dowex 50 (H^+) for the saponi-fication of ethyl acetate or for catalysis of some other suitable reaction. It is suggested that the procedure be devised for a continuous, column process rather than the processing by batches reported in the literature.

1. *Dowex 50 (H^+) as Catalyst*

To study the saponification of ethyl acetate by Dowex 50 (H^+), the following procedure may be used. Pass an aqueous solution of 0.1 M ethyl acetate through a 10 cm by 10–12 mm (I.D.) column of Dowex 50 (H^+) at room temperature. Determine the extent of hydrolysis in the effluent solution by titration of the liberated acetic acid. Then, jacket the column (Figure 8, p. 42) and repeat the experiment at higher (50–80°) and lower (0°) temperatures in order to obtain data on the extent of hydrolysis at several rates of flow and temperatures. The time of contact of ester with resin may be calculated from the flow rate and the liquid volume of the resin bed. The flow rate through the column may be readily determined by any one of several procedures that should be apparent to you. Similarly, the liquid volume of the resin bed may be determined by one of several procedures: amount of solution necessary to saturate the dry resin; the use of an acid-base indicator such as bromophenol blue as a marker of the solvent front; or the weight loss upon drying the wet resin bed in vacuo.

From this data calculate the first-order rate constant for hydrolysis at each temperature; then calculate the temperature coefficient (Q_{10}) and the heat of activation (E) for the reaction.

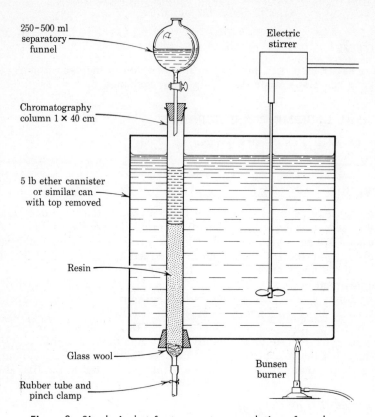

250-500 ml
separatory
funnel

Electric
stirrer

Chromatography
column 1 × 40 cm

5 lb ether cannister
or similar can
with top removed

Resin

Glass wool

Bunsen
burner

Rubber tube and
pinch clamp

Figure 8. Simple jacket for temperature regulation of a column.

2. Comparison of Dowex 50 (H⁺) and HCl as Catalysts

Determine the milli-equivalents (meq) of acid on an aliquot of Dowex 50 (H⁺) equal to that used in the above column. In order to do this, suspend the resin in distilled water, add NaCl to about 1.0 M and titrate with 0.1 N NaOH to a phenolphthalein endpoint. Mix the same number of meq of HCl in a volume equal to that of the liquid volume in the resin bed and add ethyl acetate to the 0.1 M concentration. Determine the rate of hydrolysis at the temperatures used in the resin experiment. Treat the data as above and compare the results with those obtained with Dowex 50 (H⁺).

TREATMENT OF DATA

Summarize your results and deliver a formal, oral report to the class.

QUESTIONS

1. Why was NaCl added to the suspension of Dowex 50 (H⁺) in water before titration with standard NaOH?

2. Several methods were described in the text of this experiment for the determination of the liquid volume of the resin bed. Values for this volume are not the same by these various procedures; can you suggest reasons why these discrepancies might occur?

3. Why was Dowex 50 (H$^+$) suggested for this hydrolysis and not Dowex 1 (OH$^-$)?

References

1. Khym, J. X., et al. 1953. The Identification of the Isomeric Adenylic Acids a and b. J. Amer. Chem. Soc., 75, 1262–3.
2. Sussman, S. 1946. Catalysis by Acid Regenerated Cation Exchangers. Ind. Eng. Chem., 38, 1228–30.
3. Levesque, C. L., and Craig, A. M. 1948. Kinetics of an Esterification with Cation-Exchange Resin Catalyst. Ind. Eng. Chem., 40, 96–99.
4. Thomas, G. G., and Davies, C. W. 1947. Ion Exchange Resins as Catalysts. Nature, 159, 372.
5. Bernhard, S. A., and Hammett, L. P. 1953. Specific Effects in Acid Catalysis by Ion Exchange Resins. J. Amer. Chem. Soc., 75, 1798–1800, 5834–35; ibid., 76, 991–4.
6. Whitaker, J. R., and Deatherage, F. E. 1955. Hydrolysis of Proteins and Dipeptides by Ion-Exchange Resin Catalysts. J. Amer. Chem. Soc., 77, 3360–5.

D. Filter Paper Chromatography

Filter paper chromatography is a separation process based on liquid-liquid partition of compounds on a paper support. The paper is the medium for support and retention of the non-mobile liquid phase, while the mobile liquid phase flows over the paper in intimate contact with the non-mobile phase. This extremely valuable technique was devised by Consden, Gordon, and Martin, and their classic paper (1) in 1945 set forth the method in such complete detail that the current theoretical treatment and experimental techniques remain essentially unchanged. These authors developed equations for the rate of movement of solute along the paper as a function of the distribution constant of solute between the mobile and non-mobile liquid phases, although factors such as adsorption of solute to fibers of the paper influence the rate of movement slightly. These theoretical relations have been tested and found applicable for several series of compounds, such as peptides (2) and carbohydrates (3).

Most paper chromatography depends upon a distribution of solute between an aqueous phase retained on the hydrophilic fibers of the paper and a mobile organic phase. However, separations of certain compounds such as metalloporphyrins (4) and some steroids (5) are best conducted on chromatographs in which the organic phase is the stationary phase on the paper. This is referred to as reverse phase paper chromatography and the paper is pretreated with a compound such as one of the silicones to render its fibers hydrophobic. Consult refs. (6, 7) for details.

The minimal amounts that may be separated by paper chromatography depend upon the sensitivity of methods for detection of the compounds on the paper. For amino acids, this may be as little as $0.1\ \gamma$ or 10^{-7} gm (8). With large (18″ by 22″) sheets of thick paper such as Whatman No. 3MM, many milligrams of some compounds may be separated. A modification of the preparative scale isolation of compounds by paper chromatography is the separation of compounds on columns formed from tightly wound rolls of paper held in a cylinder of polyethylene (9).

Virtually all kinds of biological compounds have been subjected to paper chromatography, and the student should consult texts (6, 7) upon the subject in order to appreciate the range of this technique. An elegant application of paper chromatography was the determination of the structure of insulin by Sanger (10). This application has been extended to studies of the amino acid sequence in peptide chains of other proteins, and a modest approach to this technique is presented in Experiment 10.

References

1. Consden, R., Gordon, A. H., and Martin, A. J. P. 1944. Qualitative Analysis of Proteins: A Partition Chromatographic Method Using Paper. Biochem. J., 38, 224–32.
2. Pardee, A. B. 1951. Calculations on Paper Chromatography of Peptides. J. Biol. Chem., 190, 757–62.

3. French, D., and Wild, G. M. 1953. Correlation of Carbohydrate Structure with Papergram Mobility. J. Am. Chem. Soc., 75, 2612—16.
4. Chu, T. C., and Chu, E. J. 1955. Paper Chromatography of Iron Complexes of Porphyrins. J. Biol. Chem., 212, 1—7.
5. Kritchevsky, T. H., and Tiselius, A. 1951. Reversed Phase Partition Chromatography of Steroids on Silicone-treated Paper. Science, 114, 299—300.
6. Lederer, E., and Lederer, M. 1953. Chromatography, A Review of Principles and Applications. Elsevier Publishing Co. New York.
7. Block, R. J., Durrum, E. L., and Zweig, G. 1955. Paper Chromatography and Paper Electrophoresis, A Manual. Academic Press. New York
8. Lakshminarayanan, K. 1954. Michrochromatography I. A Technique for Separation and Identification of Traces of Amino Acids, Sugars, Etc. Arch. Biochem and Biophys., 51, 367—70.
9. Hagdahl, L., and Danielson, C. E. 1954. A New Paper Column for Preparative Chromatography. Nature, 174, 1062—3.
10. Sanger, F., and Tuppy, H. 1951. The Amino Acid Sequence in the Phenylalanine Chain of Insulin. Biochem. J., 49, 463—90.

EXPERIMENT 10. DETERMINATION OF THE STRUCTURE
OF A PEPTIDE (2−9 periods*)

OBJECTIVES

This experiment will provide experience in the techniques of filter
paper chromatography; familiarity with certain properties of peptides
and amino acids; and some skill in manipulations at the level of milligram
amounts or less and volumes of the order of 10^{-3} milliliter.

PRINCIPAL EQUIPMENT AND SUPPLIES

Pure amino acids and di-, tri-, and tetra-peptides for student "un-
knowns"
For total hydrolysis
  Oven at 110°
  Vacuum desiccators; NaOH flakes are used to remove volatile
    acids and P_2O_5 is used for volatile bases; both will remove water
  Infra-red heating lamp
  Glass-distilled, constant-boiling HCl (20%)
  0.38 $\underline{M}$ barium hydroxide
  1.9 $\underline{M}$ sulfuric acid
For amino acid colorimetric analysis (3)
  Colorimeter to measure absorption at 570 mμ
  Water bath
  Ninhydrin solution
  80% phenol reagent
  KCN-pyridine reagent
  60% ethanol
  Standard leucine solution of 1.0 micromole (μM)/ml
For amino acid and DNP-amino acid chromatography
  Jars for ascending and descending chromatography
  Schleicher and Schuell No. 507 paper
  Whatman No. 1 paper
  Reference amino acid (Table VIII) and DNP-amino acid standards
  Ethylene chlorohydrin (glycol monochlorohydrin)
  0.8 $\underline{M}$ ammonia
  Toluene
  Pyridine
  Tertiary butanol
  Methyl ethyl ketone
  Diethylamine

---

* The length of this experiment can be adjusted to the needs of the class by the
choice of peptide "unknowns" of suitable complexity. The experiment is described
in sufficient detail to permit the student to determine the structure of a tetrapeptide; a mixture of two dipeptides; a mixture of a tripeptide and an amino acid; or
other equally complex "unknowns." The instructor should allow nine periods for the
determination of these more complex peptide unknowns.

0.1 M borate buffer, pH 9.3
Ninhydrin spray (see text for preparation)
For the DNP-peptide preparation
    Saturated NaHCO$_3$ solution
    2,4-Dinitrofluorobenzene
    Ethyl ether
    2 N KOH
For DNP-peptide hydrolysis
    Glacial acetic acid
    Perchloric acid (60%)
    Tertiary amyl alcohol
    Benzene
    2 N KOH
For partial hydrolysis
    Water bath at 55°
    0.1 N HCl, glass-distilled
    10 N HCl
    Carboxypeptidase

PROCEDURE

A. *Introduction*

The unknown may contain an amino acid or a peptide or a mixture of these compounds. The first step in the analysis of the mixture should be a test for the number of components by paper chromatography. Paper chromatography should be done with at least two solvent systems since different components may have identical R_f values in a single solvent. If a mixture is encountered, it must be separated, and further steps for its characterization will be found below. If a single peptide is present, its characterization will include (1) the identity of all amino acids present, (2) the number of moles of each amino acid per mole of peptide, and (3) the sequence of amino acids in the chain (see ref. [1]).

Diagram 1 sets forth in outline a representative approach to such a characterization. This diagram may serve also as an index of the various procedures available in this section. You need not slavishly follow the present approach. Before embarking upon a separate course, however, you should thoroughly comprehend the reason for each step and the information that may be extracted from it.

The peptide will first be totally hydrolyzed by strong acid and by strong base. All the amino acids are stable to acid hydrolysis except tryptophane and possibly tryosine. To detect the latter two amino acids, hydrolyze with alkali, although serine, threonine, cystine, and arginine are decomposed by hot alkali. In order to determine the size of the peptide, aliquots should be drawn both before and after total hydrolysis and analyzed for terminal amino nitrogen by a quantitative, colorimetric, ninhydrin method.* The amino acids present may be identified by two dimensional paper chromatography before and after hydrolysis.

---

*However, see ref. (23) in regard to this determination.

DIAGRAM 1.   STEPS IN THE ELUCIDATION OF THE STRUCTURE
OF A HYPOTHETICAL PEPTIDE, A-B-A

(Letters in parentheses refer to appropriate procedures in this experiment.)

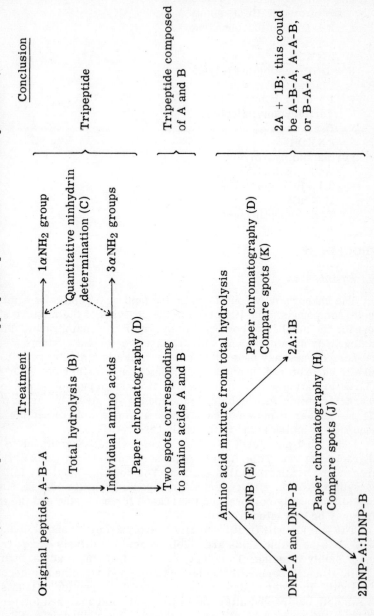

DIAGRAM 1 (Continued)

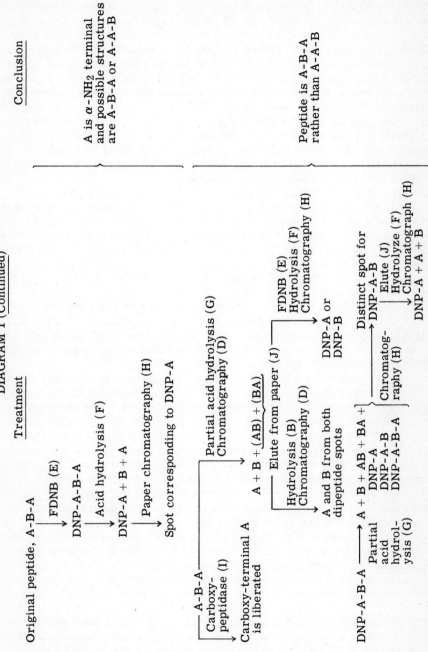

Treatment

Original peptide, A-B-A
→ FDNB (E)
DNP-A-B-A
→ Acid hydrolysis (F)
DNP-A + B + A
→ Paper chromatography (H)
Spot corresponding to DNP-A

A-B-A
→ Carboxy-peptidase (I)
Carboxy-terminal A is liberated
→ Partial acid hydrolysis (G) Chromatography (D)
A + B + (AB) + (BA)
→ Elute from paper (J)
→ Hydrolysis (B) Chromatography (D)
A and B from both dipeptide spots
→ FDNB (E) Hydrolysis (F) Chromatography (H)
DNP-A or DNP-B

DNP-A-B-A → Partial acid hydrolysis (G)
A + B + AB + BA + DNP-A DNP-A-B DNP-A-B-A
→ Chromatography (H)
Distinct spot for DNP-A-B
→ Elute (J) Hydrolyze (F) Chromatograph (H)
DNP-A + A + B

Conclusion

A is α-NH₂ terminal and possible structures are A-B-A or A-A-B

Peptide is A-B-A rather than A-A-B

The molar ratios of amino acids present may be determined by one-dimensional paper chromatography. Definite molar amounts (calculated from the ninhydrin analysis) of the peptide hydrolysate are applied to a single chromatogram sheet. Single, double, and triple molar amounts of amino acid standards will be applied to the same sheet. Comparison of intensity of spots after chromatography and treatment with ninhydrin spray may enable you to determine the molar ratios of the amino acids. If the amino acids cannot be resolved by one solvent system, another should be tried. In lieu of this procedure, a quantitative ninhydrin analysis of the amino acids extracted from the spots on paper may be made. An excellent alternate method is to react the amino acid mixture in the hydrolysate with 1-fluoro-2,4-dinitrobenzene (FDNB) and separate the 2,4-dinitrophenyl-amino acid derivatives (DNP-amino acids) by paper chromatography. The molar ratios may again be determined by elution and spectrophotometric measurement of the eluted DNP-amino acid.

The order of amino acids present in the peptide may then be determined. End group analysis with (FDNB) will identify the α-amino terminal amino acid and will completely establish the structure of a dipeptide. The DNP-peptide will be completely hydrolyzed with HCl and the DNP-derivatives may be separated from the free amino acids in the hydrolysate by extraction. Chromatography of the isolated products and comparison with suitable control compounds should completely characterize the dipeptide. For example, a dipeptide containing amino acids A and B in the order AB will yield DNP-AB. Complete hydrolysis of the derivative should yield DNP-A + B. For complete proof, DNP-A and B, but no other compounds, should be detected by chromatography. It should be noted that tyrosine, lysine, histidine, and cysteine have a second DNP-reactive group besides the free α-amino group.

The tripeptide or tetrapeptide will give somewhat more complicated results, but the terminal amino acid at the free amino end of the peptide should be determined by the DNP-derivative method.* The carboxyl-terminal amino acid may be identified by partial hydrolysis with carboxypeptidase. Alternately, partial acid hydrolysis of the peptide or the DNP-peptide, followed by chromatography of the fractions may reveal the carboxyl-terminal amino acid. For more complicated structures, elution from paper chromatograms of dipeptides formed by the above degradative procedures may be necessary. These degradation fragments may be further degraded to establish their constitution and to establish, in turn, the sequence of the amino acids in the center of the original peptide.

If the unknown is a mixture of two or more compounds, ingenuity will be required to devise the means to separate and then to determine the structure of the components, or to determine their structure without separation. A general approach to these problems will be indicated. Assume that two components appear on paper chromatography of the original unknown. If one component coincides with a known amino acid when developed in at least two solvent systems, gives the same DNP-derivative as that amino acid, and is unchanged on total acid hydrolysis, its identity is established. If the second component does not coincide with a known amino acid and is changed on total acid hydrolysis, it is a peptide. The

*Alternatively, the phenylisothiocyanate method (1) may be used.

structure of the peptide can be determined in the presence of the other component, the amino acid, by the procedures outlined above. In the event that both compounds are peptides, separation and determination of the structure of the separated peptides is necessary. The separation may be effected by formation of the DNP-derivatives of the components of the mixture. The DNP-derivatives may be separated by paper chromatography and the separate components eluted from the paper. The separate DNP-derivatives may then be hydrolyzed and again chromatographed to determine the amino acid composition and sequence of each component.

B. *Total Hydrolysis of the Peptide*

The following procedure is that of Levy and Chung (2). Approximately 1 mg of the peptide is weighed into a 2 in. length of Pyrex tubing (4 mm I.D.) which has been sealed at one end, and 0.1 ml of glass-distilled constant boiling (20%) hydrochloric acid is added by pipet. (Undistilled hydrochloric acid may contain trace metals that catalyze humin formation.) The bottom of the tube is chilled in ice and the tube is evacuated and then sealed. The sealed tube is heated in an oven at 110° for 6 hr; longer periods of hydrolysis may cause a loss of serine and threonine. Then the tube is opened and the acid is evaporated over NaOH flakes in a vacuum desiccator. This requires 0.5 – 1.0 hr if the desiccator is cautiously heated with an infra-red heat lamp. The dried residue is dissolved in 1 ml of water and again taken to dryness to insure removal of HCl.

Alkaline hydrolysis is carried out by heating approximately 1 mg of the peptide with 100 μl (0.1 ml) of 0.38 $\underline{M}$ barium hydroxide at 100° for 3 hr in a sealed tube. 20 μl of 1.9 $\underline{M}$ sulfuric acid is added to the cooled tube, the precipitated barium sulfate is removed by centrifugation, and the supernatant solution is evaporated to dryness as above.

C. *Colorimetric Amino Nitrogen Determinations*

Caution: the room should be free of ammonia fumes and tobacco smoke.

Aliquots of the amino acid, peptide, or peptide hydrolysate that contain 0.05 to 0.5 micromole of amino acid at any pH between 1 and 8 are pipetted into test tubes for α-amino nitrogen analysis by the quantitative ninhydrin method of Troll and Cannan (3). Various volumes of 1.0 millimolar (mM) leucine are pipetted into other tubes to provide standards, and 0.5 ml of distilled water is pipetted into another tube as a reagent blank. All volumes are adjusted to 0.5 ml by proper additions of distilled water, and 1.0 ml of KCN-pyridine reagent and 1.0 ml of 80% phenol reagent are added. (Caution: draw these reagents into the pipet by means of a suction bulb.) Thoroughly mix the solutions and cap the tubes with aluminum foil. Place the tubes in a boiling-water bath and, when the tube contents have reached 98°, add 0.2 ml of ninhydrin solution to each tube. (Again pipet with a suction bulb.) Leave the tubes in the bath for an additional 5 min; then cool them under a cold water tap. Make up the volume of each tube to 10 ml with 60% ethanol and determine the optical density at 570 mμ.

The colorimeter should be adjusted to the reagent blank for zero color intensity. A standard curve is constructed and the free α-amino nitrogen content of the unknown is determined from that curve. Color values are equivalent ($100 \pm 2\%$) on a molar basis to that of leucine for the common amino acids, except tryptophane (75% as much color) and lysine (110%). Ammonia will react in this assay and the color is equivalent to 29% of that obtained with leucine.

D. *Qualitative Amino Acid Identification*

Identification of the component amino acids in the hydrolysis mixture obtained in part B of the experiment may be made by paper chromatography. Initially a two-dimensional chromatogram should be prepared. Maximum resolution of all amino acids in the mixture is thereby obtained. Subsequently, one-dimensional chromatograms may be necessary to complete the identification.

Two-dimensional chromatography (4)

Solvent I		Solvent II	
Methyl alcohol	80 vol.	Tertiary butyl alcohol	40 vol.
H_2O	20 vol.	Methyl ethyl ketone	40 vol.
Pyridine	4 vol.	H_2O	20 vol.
		Diethylamine	4 vol.

The sample to be chromatographed is placed on a spot in the lower left-hand corner and 2 cm from each edge of squares (15 by 15 cm) of Schleicher and Schuell No. 507 paper. Applied amounts should make a spot 3 mm or smaller in diameter and 0.01 millimoles of an amino acid is sufficient. The solutions may be applied to the paper conveniently by means of capillary tubes or micropipets. After the applied spot is dry, staple the paper in the form of a cylinder but do not allow edges of the paper to come in contact. Place the cylinder in a screw-cap jar (approx. 18 cm high by 9 cm diameter) that contains about 20 ml of solvent I in the bottom and that has been stored several hours for equilibration of vapors. Time for chromatogram development will be 2–3 hr. The paper is dried in a hood for 30 min, the staples are removed, and a new cylinder is formed by rotating the paper 90° so that the base of the cylinder now will contain the strip of amino acids developed in the first dimension. The chromatogram is then developed as above in a second jar with solvent II. Time of development is 3–5 hr. The cylinder is again dried in the hood.

Before being treated with ninhydrin, the paper is sprayed with 0.1 M borate buffer, pH 9.3, and dried over a hot plate to eliminate background color from diethylamine. The amino acids are located by spraying the paper with a mixture of ninhydrin, collidine, and glacial acetic acid (50 ml of 0.1% ninhydrin in ethanol plus 2 ml of collidine and 15 ml of glacial acetic acid). While the paper is still wet, it is held 2–3 in. above a hot plate until colored spots due to the amino acids are completely developed. Note the initial colors of the spots, for frequently these fade or change on prolonged heating.

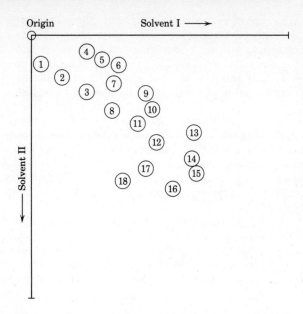

Figure 9. Two-dimensional paper chromatography of amino acids by the procedure of Redfield.

1. Cystine	7. Glycine	13. Valine
2. Lysine	8. Serine	14. Isoleucine
3. Histidine	9. Alanine	15. Leucine
4. Arginine	10. Proline	16. Phenylalanine
5. Aspartic acid	11. Tyrosine	17. Threonine
6. Glutamic acid	12. Methionine	18. Tryptophane

TABLE VIII. REFERENCE AMINO ACID MIXTURES
FOR PAPER CHROMATOGRAPHY

Standard A	Color of Ninhydrin Spot*	Standard B	Color of Ninhydrin Spot*
Lysine	blue — purple	Cystine	brown
Aspartic acid	blue	Histidine	turquoise blue
Glycine	reddish purple	Arginine	blue — purple
Threonine	grey	Serine	muddy yellow
Proline	yellow	Glutamic acid	blue — purple
Valine	blue — purple	Alanine	blue — purple
Tryptophane	yellowish brown	Tyrosine	turquoise blue
Phenylalanine	turquoise blue	Methionine	grey
Leucine	blue — purple	Isoleucine	blue — purple

*Initial color; colors fade rapidly.

A chromatogram of reference amino acid standards A and B should be developed similarly on a separate sheet. This standard may be compared, for identity of spots, with the chromatogram from the peptide hydrolysate. The chromatogram obtained with the reference standards should be similar to Figure 9. The amino acids that should be available as standard solutions of 20 micromoles of each amino acid per ml are listed in Table VIII.

One-dimensional chromatography in descent. In the event that the above procedure does not decisively distinguish among several possible amino acids, definitive identification may be obtained by one-dimensional chromatography. Choose the solvent system most likely to distinguish the various possible amino acids. Consult references (5) and (6) for additional solvents. Leucine and isoleucine may be distinguished by a solvent system given in reference (1).

Cut $4\frac{1}{2}$-in. by 16-in. strips of Whatman No. 1 paper. Rule 1 in. lanes lengthwise of the strip starting $\frac{1}{2}$ in. from one edge. Perpendicular to these, rule a line $1\frac{1}{2}$ in. from the top of the strip as shown in Figure 10. The hydrolysate or standard amino acid is applied at a spot where vertical and horizontal lines cross. Applications should be of the order of 2 μl and the

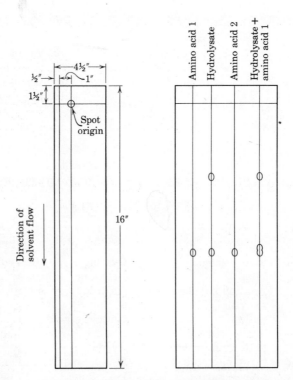

Figure 10. Arrangement of paper for paper chromatography in descent.
Figure 11. Developed chromatogram.

area of the spot should be 5 mm or less. Multiple applications may be necessary in order to get an optimal amount (about 0.05 micromoles) on the spot. On the same sheet place spots of the hydrolysate and of amino acids that might be identical with those in the hydrolysate. An additional, excellent procedure is the application of unknown and known amino acids on the same spot, for non-identity may be revealed even though differences in movement are slight (Figure 11). An attempt should be made to estimate the ratio of the number of moles of each amino acid present on the basis of area and density of ninhydrin spots. If the results are inconclusive, use the quantitative method given below.

E. *Preparation of Dinitrophenyl Derivatives* (7 - 10)

First, 2 − 3 mg of the peptide or amino acid is dissolved in 0.5 ml of saturated $NaHCO_3$ solution. To this mixture is added 10 mg of 1-fluoro-2,4-dinitrobenzene (FDNB) dissolved in 0.5 ml of ethanol. The mixture is shaken at room temperature for 2 hr, concentrated in vacuo to remove ethanol, dissolved in water, and extracted several times with ether to remove excess FDNB. The aqueous layer is acidified with HCl to pH 2, whereupon the DNP-derivative usually separates as an oil. The DNP-derivative then can be extracted into ether or, preferably, ethyl acetate for DNP-peptides. Crystallization is usually unnecessary but can be done from methanol or aqueous ethanol if desired. The acid-soluble DNP-amino acids are mono- and di-DNP-histidine,* S-DNP-cysteine, α-DNP-arginine, ϵ-DNP-lysine, O-DNP-tyrosine (colorless), and α-DNP-lysine.* All other DNP-amino acids are soluble in ether after acidification, including the di-DNP-derivatives of lysine, tyrosine, and cysteine. 2,4-Dinitrophenol is usually produced as a hydrolysis product of FDNB and DNP-derivatives and may interfere with chromatographic identification, particularly of DNP-alanine (Table X). However, the 2,4-dinitrophenol can readily be identified by its spectrum (Table XI) and by the fact that its color disappears on acidification. 2,4-Dinitroaniline also is produced, particularly in the presence of light. Standard DNP-derivatives should be stored in the form of dry salts in the refrigerator and taken up in a little methanol prior to chromatography.

F. *Hydrolysis of the DNP-peptide*

Most DNP-peptides may be hydrolyzed by HCl, using the same procedure indicated for the original peptide (see part B of this experiment). Some DNP-derivatives are broken down by hot HCl (Table IX) and the following modified procedure of Hanes et al. (11) should be used. The DNP-peptide (2 mg of original peptide) is dissolved in 0.1 ml of a mixture of glacial acetic acid (9 vol.) + 60% (w/v) perchloric acid (1 vol.), and the mixture is heated for 8 hr at 100° in a sealed capillary. The hydrolysate is then diluted with 1 ml of water and extracted three times by shaking with equal volumes of 10% (v/v) tertiary amyl alcohol in benzene for the

---

*Questionable; depends on efficiency of extraction.

TABLE IX. APPROXIMATE BREAKDOWN OF DNP-AMINO ACIDS ON ACID HYDROLYSIS * (7)

Compound †	Per Cent Breakdown
Bis-DNP-cystine	100
DNP-glycine	50
DNP-hydroxyproline	50
DNP-phenylalanine	50
DNP-proline	50
DNP-tryptophane	100
DNP-tyrosine	50

* 16 hours with 12 N HCl at 105°.

† DNP-amino acids not included are altered to the extent of 25% or less under these conditions.

acetic-perchloric hydrolysate, or with ether for the HCl hydrolysate. The DNP-derivatives are recovered for chromatographic identification from the organic extracts by evaporation to dryness. Traces of acid that remain in the ether residue should be completely removed in vacuo. Hydrolysis products that remain in the original aqueous HCl mixture also are recovered for chromatographic identification by evaporation to dryness over NaOH flakes. Hydrolysis products that remain in the original aqueous acetic-perchloric mixture are recovered for chromatographic identification by neutralization of the mixture to pH 3 with 2 N KOH and evaporation of the solution to dryness in vacuo over NaOH flakes. The residue is thoroughly extracted with 50% (v/v) aqueous ethanol to dissolve free amino acids, and the extracts are filtered from $KClO_4$ and concentrated for chromatographic identification.

G. *Partial Acid Hydrolysis of the Peptide or the DNP-peptide*

Partial hydrolysis with acids is achieved by reducing the time of hydrolysis, by decreasing the concentration of the acid, or by operating at lower temperatures. Gordon, Martin, and Synge (12) have cited evidence that hot dilute acids or enzymes might cause rearrangement or ring formation; hence, these workers preferred using strong mineral acid in the cold. This method was used in elaborating the structure of Gramicidin S (13), and Sanger (14) used a similar procedure to obtain peptides from the phenylalanine chain of insulin.

Dipeptides have a greater stability in acid hydrolysis than higher peptides, and they are the major constituents after hydrolysis. The bonds involving amino groups of serine and threonine are very labile; bonds involving glycine are somewhat less labile; and glycine peptides appear relatively more stable to dilute mineral acid at elevated temperature than to strong mineral acid in the cold. Partial acid hydrolysis of the DNP-peptide is sometimes preferable to hydrolysis of the original peptide. Hydrolysis with alkali results in considerable destruction of many amino acids even while in peptide linkage, and it is not suitable for studies of partial hydrolysis.

Procedure with concentrated mineral acid. 1.0 mg of peptide or DNP-peptide is mixed with 100 μl of 10 N HCl and kept at 55°. Aliquots are drawn for chromatography in the interval 1–48 hr. The aliquot may be applied directly to filter paper for chromatography, but the spot should be dried thoroughly to expel all HCl before developing the chromatogram. If tryptophane is present, partial hydrolysis is done in a sealed tube under nitrogen. If DNP-amino acids sensitive to HCl are present, the perchloric-acetic acid mixture of Hanes (see part F) should be used.

Procedure with dilute mineral acid. 1.0 mg of peptide is mixed with 100 μl of 0.1 N glass-distilled HCl and stored at 100°. Draw aliquots at 24-hr intervals and test them as indicated for the products of hydrolysis by concentrated acid.

H. *Chromatography of DNP-derivatives*

The DNP-amino acids or DNP-peptides may be separated by the paper chromatography method of Biserte (15). Alternatively, the procedure of Blackburn and Lowther (16) may be followed. The solvent system of Biserte is prepared by the addition of 50 ml of toluene, 10 ml of pyridine, and 30 ml of ethylene chlorohydrin to a separatory funnel.* The solvents are mixed by gentle swirling and 30 ml of 0.8 M ammonia is carefully introduced drop by drop down the side of the separatory funnel. The two phases are allowed to equilibrate undisturbed for at least 1 hr before the aqueous layer is carefully drained off and saved. The organic phase should now be perfectly clear and colorless; cloudiness indicates that insufficient time was taken for equilibration. The organic phase is next filtered through dry Whatman No. 1 filter paper to remove suspended drops of water.

The aqueous phase and approximately 50 ml of 0.8 M ammonia are poured into the bottom of a chromatography jar designed for solvent flow in descent. The solvents and vapor phase should be allowed to equilibrate in the closed jar for 2 to 3 hr prior to introduction of the chromatogram. Meanwhile, spots of the compounds are applied to Whatman No. 1 paper. A moderately yellow spot should be applied and the spot diameter should not exceed 5 mm, for the paper is easily overloaded. The prepared paper sheet is stored in the jar at least 1 hr before the organic phase is added to the trough. Development of the chromatogram will require 6–12 hr, and the jar must be kept sealed and protected from temperature variations.

Spots of the individual DNP-amino acids or DNP-peptides may be eluted from the paper by 1% NaHCO$_3$ and their identities and relative amounts may be determined by spectrophotometric methods. The ratio of absorbancy at 390 and 360 mμ partially serves to characterize the DNP-derivative (Table XI); and the amount of the DNP-amino acid eluted from the paper may be estimated from the absorbancy at 360 mμ and an approximate molar extinction coefficient of 17,000 at 360 mμ.

---

* The best commercial grades of these solvents should be obtained; then no additional purification is required.

TABLE X. R_f VALUES OF DNP-AMINO ACIDS
IN THE SOLVENT SYSTEM OF BISERTE (15)

DNP-amino acids	R_f	DNP-amino acids	R_f
N-DNP-cysteine	0.02	2,4-Dinitrophenol	0.42 – 0.45
$\overline{\text{D}}$NP-aspartic acid	0.08	DNP-valine	0.56
DNP-glutamic acid	0.09	DNP-methionine	0.59
DNP-asparagine	0.24	DNP-tryptophane	0.62
DNP-glutamine	0.27	DNP-leucine	0.64
DNP-hydroxyproline	0.28	DNP-phenylalanine	0.66
DNP-serine	0.30	Di-DNP-lysine	0.72
Bis-DNP-cystine	0.32	Di-DNP-tyrosine	0.74
DNP-glycine	0.35	2,4-Dinitroaniline	0.95 – 1.0
DNP-threonine	0.36	ϵ-DNP-lysine	0.40
DNP-proline	0.43	α-DNP-arginine	0.48
DNP-alanine	0.45	Di-DNP-histidine	0.34; 0.70

TABLE XI. ABSORPTION CHARACTERISTICS
OF DIFFERENT DNP COMPOUNDS (17)

	Ratio of O.D. at 390/360 mμ	Approximate Maxima
DNP-proline	1.40	390
DNP-prolyl peptides	1.05	385
DNP-hydroxyproline	1.20	
Dinitrophenol	0.80	
Other DNP-amino acids	0.60 – 0.65	360
Di-DNP-ornithine	0.55 – 0.60	
Di-DNP-lysine	0.55 – 0.60	
Di-DNP-tyrosine	0.55 – 0.60	
Di-DNP-cystine	0.55 – 0.60	
Other DNP-peptides	0.50 – 0.55	355

I. *Enzymatic Hydrolysis of the Peptide or DNP-peptide with Carboxypeptidase* (1)

Enzymatic hydrolysis by carboxypeptidase offers considerable advantage over partial hydrolysis by acid. The carboxypeptidase specifically hydrolyzes the carboxyl-terminal amino acid. Since the enzyme seldom attacks dipeptides, only one amino acid will be liberated from a tripeptide; two may appear after exhaustive treatment of a tetrapeptide. It has been found that carboxypeptidase will in some cases hydrolyze the free peptide and not the DNP-derivative, while in others the DNP-derivative is cleaved; therefore one should subject both the free peptide and the DNP-derivative to the action of carboxypeptidase. To prepare the carboxypeptidase solution, suspend 1 mg of crystalline carboxypeptidase in 1 ml of 0.1% NaHCO$_3$ at 0 – 4° and dissolve the crystals by addition with stirring of 0.05 – 0.10 ml of $0.1 \underline{N}$ NaOH. The solution is quickly adjusted to pH 8 by the addition of $0.1 \underline{N} \overline{HCl}$ (about 0.1 ml). Dissolve the peptide sample in 0.1 ml of distilled water and add 0.1 ml of the carboxypeptidase solution. Incubate at 25° and remove samples at intervals, i.e., 0, 1, 2, 4, 8, 16, 32 hr. These samples are spotted on paper and then developed in one dimension, using the ascending technique and Redfield's solvent I. The amino acids suspected of being carboxy terminal should also be spotted on the paper to compare with the amino acid liberated by the action of carboxypeptidase.

J. *Elution of Spots from Paper Chromatograms* (18)

Paper chromatography can be used for the isolation of micro-quantities of amino acids and peptides as well as for their separation and detection. It may be necessary, for example, to isolate a dipeptide obtained from a partial hydrolysate of a larger peptide. If the spot is fluorescent in ultraviolet light, it can be located and cut out. Heating the paper at 100° for 15 min may make many amino acids visible under ultraviolet light if phenol is not present. Otherwise, it is necessary to chromatograph duplicate sets of spots on the same sheet. One-half the sheet is sprayed with ninhydrin to locate the spots, and the corresponding spots of the duplicate set are cut from the second half of the sheet.

A spot should be cut out so that it is located at the top of a short strip which is pointed at the lower end. The wide end of the strip may be dipped into water in a regular chromatography jar and the eluent that drops from the pointed tip may be collected in a small test tube. Alternately, the micro procedure of Sanger (14) may be followed. The paper strip is placed in a very small paper chromatography apparatus, and the eluent that drips from the strip is collected in a small tube. The eluent then is evaporated in vacuo to dryness in the small tube. In each of these procedures, elution must be done in a jar in which the atmosphere is saturated with the solvent for elution (usually water).

A simpler procedure for removal of the compound from the paper spot is direct extraction of the paper with the solvent and subsequent separation of the extract from the paper by filtration, or by centrifugation and decantation.

K. *Quantitative Amino Acid Analysis on Paper Chromatograms* (18–22)

A single dimension paper chromatogram that will resolve the amino acids present in the peptide is prepared as in part D. The room should be free of ammonia and tobacco smoke, and at no time should the sheet be handled with the fingers. Utmost cleanliness is essential. Apply 0.04, 0.08, and 0.16 micromoles of standard amino acids in 2 μl applications in duplicate, 1 in. apart at the origin. Also apply 0.12 and 0.24 micromoles of amino nitrogen in the peptide hydrolysate in duplicate. The final developed spot area must in all cases fit into an area less than 1 in.^{2}. The sheet is developed as usual, taking care that the solvent front moves parallel to the origin line. The solvent front is marked, and the sheet is dried in the hood. The first lane of each duplicate sample is carefully cut out and sprayed with ninhydrin. These spots are used to locate the amino acids on the unsprayed lanes. The spots are carefully cut out as 1 in. squares (handle with clean forceps), folded and pushed into test tubes. Three tubes are prepared as blanks; they contain 1 in. squares cut from regions of the chromatogram where no spots appear. Two other blank tubes that contain no paper are prepared. The paper in each tube is uniformly moistened with 0.5 ml of 0.1 N NaOH, and the tubes are stored overnight in a vacuum desiccator over H_2SO_4 in order to remove residual ammonia from the paper. Alternatively, the ammonia may be removed by placing the tubes containing alkali in a boiling-water bath, and bubbling acid-washed nitrogen through them for 2 min. After either procedure for removal of ammonia, 0.5 ml of 0.1 N HCl is added to each tube followed by the other reagents for the quantitative α-amino nitrogen analysis as described in part C. Samples may be centrifuged or may be filtered through ammonia-free paper before measurement of the color intensity. A standard curve may be drawn from the samples of the known amino acids and the amounts of amino acids in the peptide hydrolysate may be estimated. With good technique this method should be accurate to at least ± 10%.

TREATMENT OF DATA

No written report is necessary, but you will summarize in the form of an outline the steps taken to characterize the peptide unknown. The summary must be supplemented by chromatograms and other pertinent data, and you should discuss this evidence with the instructor.

References

1. Fraenkel-Conrat, H., Harris, J. I., and Levy, A. L. 1955. Recent Developments in Techniques for Terminal and Sequence Studies in Peptides and Proteins. In Methods of Biochemical Analysis, II, 360—420. D. Glick, editor. Interscience Publishers. New York.
2. Levy, A. L., and Chung, D. 1953. Two-dimensional Chromatography of Amino Acids on Buffered Papers. Anal. Chem., 25, 396—9.
3. Troll, W., and Cannan, R. K. 1953. A Modified Photometric Ninhydrin Method for the Analysis of Amino and Imino Acids. J. Biol. Chem., 200, 803—11.
4. Redfield, R. R. 1953. Two-Dimensional Paper Chromatographic Systems with High Resolving Power for Amino Acids. Biochem. et Biophys. Acta, 10, 344—5.
5. Hardy, T. L., Holland, D. O., and Nayler, J. H. C. 1955. One-phase Solvent Mixtures for the Separation of Amino Acids. Anal. Chem., 27, 971—4.
6. Clayton, R. A., and Strong, F. M. 1954. New Solvent System for Separation of Amino Acids by Paper Chromatography. Anal. Chem., 26, 1362—3.
7. Porter, R. R. 1950. Use of 1:2:4-Fluorodinitrobenzene in Studies of Protein Structure. In Methods in Medical Research, III, 256—71. Year Book Publishers. Chicago, Ill.
8. Schroeder, W. A., and LeGette, J. 1953. A Study of the Quantitative Dinitrophenylation of Amino Acids and Peptides. J. Amer. Chem. Soc., 75, 4612—5.

9. Rao, K. R., and Sober, H. A. 1954. Preparation and Properties of 2,4-Dinitro-phenyl-L-Amino Acids. J. Amer. Chem. Soc., 75, 1328–31.
10. Levy, A. L., and Chung, D. 1955. A Simplified Procedure for the Synthesis of 2,4-Dinitrophenyl-Amino Acids. J. Amer. Chem. Soc., 77, 2899–2900.
11. Hanes, C. S., Hird, F. J. R., and Isherwood, F. A. 1952. Enzymatic Transpepti-dation Reactions Involving γ-Glutamyl Peptides and α-Amino-Acyl Peptides. Biochem. J., 51, 25–35.
12. Gordon, A. H., Martin, A. J. P., and Synge, R. L. M. 1941. A Study of the Partial Acid Hydrolysis of Some Proteins, with Special Reference to the Mode of Linkage of the Basic Amino Acids. Biochem. J., 35, 1369–87.
13. Consden, R., Gordon, A. H., and Martin, A. J. P. 1947. The Sequence of the Gramicidin S Residues. Biochem. J., 41, 596–602.
14. Sanger, F., and Tuppy, H. 1951. The Amino Acid Sequence in the Phenylala-nine Chain of Insulin. Biochem. J., 49, 463–90.
15. Biserte, G., and Osteux, R. 1951. Separation of DNP-Amino Acids on Paper Chromatograms. Bull. Soc. Chimie Biol., 33, 50–63.
16. Blackburn, S., and Lowther, A. G. 1951. Separation of N-2:4-Dinitrophenyl Amino Acids on Paper Chromatograms. Biochem. J., 48, 126–8.
17. Niu, C. I., and Fraenkel-Conrat, H. 1955. Determination of C-Terminal Amino Acids and Peptides by Hydrazinolysis. J. Am. Chem. Soc., 77, 5882–5.
18. Block, R. J., Durrum, E. L., and Zweig, G. 1955. Paper Chromatography and Paper Electrophoresis, A Manual. Academic Press. New York.
19. Fowden, L. 1951. The Quantitative Recovery and Colorimetric Estimation of Amino Acids Separated by Paper Chromatography. Biochem. J., 48, 327–33.
20. Isherwood, F. A., and Cruickshank, D. H. 1954. A New Method for the Colori-metric Estimation of Amino Acids on Paper Chromatograms. Nature, 174, 123–6.
21. Levy, A. L. 1954. A Paper Chromatographic Method for the Quantitative Esti-mation of Amino Acids. Nature, 174, 126–7.
22. Kay, R. E., Harris, D. C., and Entenman, C. 1956. Quantification of the Nin-hydrin Color Reaction as Applied to Paper Chromatography. Arch. Biochem. and Biophys., 63, 14–25.
23. Yanari, S. 1956. The Reaction of Ninhydrin with Dipeptides: Differences in Reaction Rates and Theoretical Yield. J. Biol. Chem., 200, 683–9.

EXPERIMENT 11. DEVELOPMENT OF A PAPER CHROMATOGRAPHIC METHOD (3– 4 periods)

OBJECTIVE

This experiment will extend your knowledge of paper chromatography to an understanding of how a paper chromatography method may be developed.

PRINCIPAL EQUIPMENT AND SUPPLIES

A variety of large and small tubes and jars suitable for ascending and descending paper chromatography in one and two dimensions
A wide range of solvents and grades of filter paper

PROCEDURE

The instructor will assign a problem for the separation of several chemically related compounds, or the student may suggest a separation of interest to him. In any case, the student is ''honor bound'' not to refer to the literature on an identical separation. Reference to the separation of chemically similar compounds is acceptable; in fact, it is a valuable lead to suitable experimental conditions. It is suggested that an initial survey of solvents, grades of paper, pH, temperature, etc., be made by the rapid test-tube scale technique of Rockland et al. (1, 2). Refinement of conditions and determination of R_f values then may be made in larger apparatus.

TREATMENT OF DATA

Summarize your results and deliver a formal, oral report to the class.

References

1. Rockland, L. B., and Dunn, M. S. 1949. A Capillary-Ascent Test Tube Method for Separating Amino Acids by Filter Paper Chromatography. Science, 109, 539–40.
2. Rockland, L. B., and Underwood, J. C. 1954. Small-Scale Filter Paper Chromatography. Anal. Chem., 26, 1557–63.

E. Zone Electrophoresis

The two general techniques for the study of ion migration in solution are boundary electrophoresis and zone electrophoresis. In boundary electrophoresis, ions in solution migrate in an electric field at rates which are primarily a function of the ionic charge. Because of density considerations to be discussed later, boundary electrophoresis experiments are generally performed in U-tubes, and the components of a mixture will separate during this process only to the extent illustrated in Figure 12. The individual components cannot be completely separated into distinct zones because zones of greater density (solvent plus solute) cannot be maintained above those of less density (solvent alone); rather, the more dense zones would flow back into those less dense. This convection would mix the solutions and destroy any zones as they form. However, this difficulty is overcome in the second general technique, zone electrophoresis.

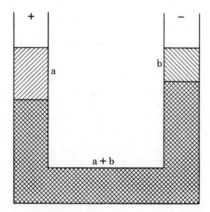

Figure 12. Schematic boundary electrophoresis tube.

Zone electrophoresis is a process of ion migration in an electric field in solutions that are stabilized from convection by some supporting medium. Because of the freedom from convection, ions of different mobilities travel as separate zones (1, 2). One way to stabilize the solution against convection is to superimpose throughout it a uniform density gradient which is greater than those created by the zones of migrating ions (Figure 13). For example, zone electrophoresis has been conducted successfully in solutions with density gradients that were created by varying the concentration of an uncharged solute, such as sucrose (4, 5). Zone electrophoresis in solutions with a density gradient has the advantages that there are none of the complicating factors of adsorption or electro-osmosis that occur in other types of zone electrophoresis discussed below, and that the migration of zones may be followed by optical means.

Solid supporting media, however, are the most widely employed for zone electrophoresis; here convection is prevented and zones of different density can form because the solution exists in capillary passages of the

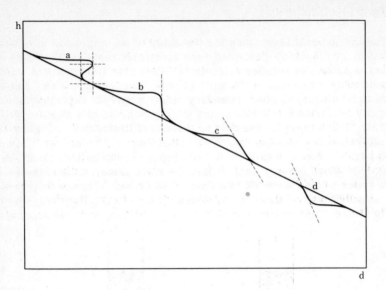

Figure 13. Density (d)—Height(h) diagram. Solid straight line: density course of solvent. Solid curves: density courses through solute zones. (a) Unstable zone: convection will occur within the dotted rectangle. (b) Stable but sensitive zone with a gradient-free portion where the curve tangent is vertical. (c) Stable zone: curve tangent is never vertical. (d) Stable zone with a negative density increment.

solid medium. Paper is the most popular supporting medium, and paper electrophoresis rivals paper chromatography in diversity of apparatus and range of applications (6). Starch gel (7) is a newer medium which appears to be less subject to the complications mentioned below for solid supporting media. Other supporting media are starch, sponge rubber, silica gel, asbestos, etc. (8). The latter materials are treated with the buffer and held in troughs during electrophoresis. With such apparatus, milligram amounts of compounds can be separated, eluted from the supporting medium, and recovered.

Zone electrophoresis has certain advantages over boundary electrophoresis: the complete separation into zones permits isolation of separated compounds, boundary anomalies interfere less, and the method can be adapted to very small quantities. However, some complications are introduced by the presence of supporting medium and render zone electrophoresis less subject to accurate theoretical treatment than boundary electrophoresis. That is to say, zone electrophoresis in paper or other solid supporting media is not simply a process of ion migration in a uniform electric field. Frequently the electric field is not uniform and the buffer concentration also may vary because of evaporation of solvent. Adsorption of solutes by the supporting medium may occur, particularly with proteins as solutes. Electro-osmotic flow of solvent is still another complicating factor. Solvent flow occurs because differences in electrical charge exist between the solvent and the supporting medium, and, as a consequence, the solvent is impelled toward the electrode of opposite charge. Therefore, uncharged molecules move with the solvent during electrophoresis and charged molecules move at rates different from those that would occur in the absence of electro-osmosis. Electro-osmotic flow should always be taken into account in zone electrophoresis and it may be determined approximately by the migration of an uncharged substance during the electrophoresis. A compound that is readily detected should be used for this purpose; for example, chloromycetin is used in Experiment 12 because the compound is readily located under an ultraviolet lamp.

References

1. Tiselius, A., and Flodin, P. 1953. Zone Electrophoresis. In Advances in Protein Chemistry, VIII, 461–85. Academic Press. New York.
2. Kunkel, H. G. 1954. Zone Electrophoresis. In Methods of Biochemical Analysis, I, 141–70. D. Glick, editor. Interscience Publishers. New York.
3. Svensson, H., and Valmet, E. 1955. Density Gradient Electrophoresis—A New Method of Separating Electrically Charged Compounds. Science Tools, 2, 11–13.
4. Kolin, A. 1954. Separation and Concentration of Proteins in a pH Field Combined with an Electric Field. J. Chem. Phys., 22, 1628–9; Proc. Natl. Acad. Sci. U. S., 41, 101–10.
5. Brakke, M. K. 1955. Zone Electrophoresis of Dyes, Proteins and Viruses in Density Gradient Columns of Sucrose Solutions. Arch. Biochem. and Biophys., 55, 175–90.
6. Block, R. J., Durrum, E. L., and Zweig, G. 1955. Paper Chromatography and Paper Electrophoresis, A Manual. Academic Press. New York.
7. Bernfeld, P., and Nisselbaum, J. S. 1956. A Modified Method for Protein Separation by Zone Electrophoresis on A Starch Gel. J. Biol. Chem., 220, 851–60.
8. MacDonald, H. J. 1955. Ionography, Electrophoresis in Stabilized Media. Year Book Publishers. Chicago, Ill.

EXPERIMENT 12.   ZONE ELECTROPHORESIS ON PAPER (3 periods)

OBJECTIVE

This experiment will serve to introduce the technique of paper electrophoresis.

PRINCIPAL EQUIPMENT AND SUPPLIES

Adjustable power source; 0–500 volts and 0–200 ma

Paper electrophoresis chamber (see ref. [1] for types of apparatus available)

Whatman No. 3MM paper

Snake venom, Crotalus adamantus*

Bis(p-nitrophenol)phosphate (Ca salt)†

Barbiturate buffer; 0.05 M, pH 9.2

Protein dye solution (0.1 g bromphenol blue and 50 g $ZnSO_4 \cdot 7H_2O$ diluted to 1 liter with 5% acetic acid)

Protein dye fixative (3.0 g sodium acetate·$3H_2O$ diluted to 1 liter with 5% acetic acid)

Chloromycetin (0.3 mg/ml)

PROCEDURE

The separation in 1939 of the protein components of snake venom was one of the earliest applications of paper electrophoresis (3). Among the components of venom is the enzyme phosphodiesterase (4). In the present experiment, the phosphodiesterase will be separated from the other components of the venom by paper electrophoresis, and both the percentage of recovery of the enzyme and the change in specific activity are to be determined.

Dissolve 1 mg of the dried venom in 1.0 ml of water to form the stock enzyme solution. Analyze this solution for protein content by the method given in Appendix II, and for phosphodiesterase activity by the assay of Sinsheimer and Koerner (5).

Assemble the electrophoresis apparatus‡ and equilibrate Whatman No. 3MM paper strips with 0.05 M barbiturate buffer of pH 9.2. Apply the enzyme solution in amounts of 20 λ (1 λ = 0.001 ml) per 1 in. strip to five strips of the paper and to the fifth strip also apply 20 λ of chloromycetin (0.3 mg/ml) as a marker of electro-osmosis. (Caffeine may be used similarly for this purpose.) The materials should be applied to the paper along a line that is equidistant from the two electrode vessels. Run the electrophoresis in a cold room (4–10°) at 220 v for 16 hr; then dry the papers in a hood at room temperature.

---

*Ross Allen Reptile Institute, Silver Springs, Florida.

†Prepared from tris(p-nitrophenol)phosphate (Aldrich Chemical Co., Milwaukee, Wisconsin) by the method of Corby, et al. (2).

‡Conditions for this experiment were established for the Spinco paper electrophoresis apparatus, Model R. See ref. (6).

One strip should be stained for protein with bromphenol blue by the method described in Chapter XVIII of ref. (1). The relative protein content of the various bands may be estimated by scanning the dyed strip with a densitometer. Alternatively, the dyed strip may be used as a guide for location of corresponding zones on a second strip; and the proteins may be eluted from the second strip (although not quantitatively) and measured by the method described in Appendix II.

The phosphodiesterase activity of the various bands may be estimated in one of two ways, both of which are based on the use of bis(p-nitrophenol)phosphate as substrate (5) according to the following equation:

| Colorless | Colorless | Yellow |

In one procedure, a third strip from the electrophoresis is placed on a glass plate and sprayed with a 2 $\underline{M}$ solution of bis(p-nitrophenol)phosphate at pH 8.8. The strip should be sprayed so that the paper is uniformly but lightly moistened. The strip on the glass plate is then incubated at room temperature in an atmosphere saturated with water vapor so that the paper does not become dry. The phosphodiesterase zone should appear as a yellow band within several hours.

In the second procedure, the first strip, dyed with bromphenol blue, is again employed as a guide and corresponding zones are cut from the fourth electrophoresis strip. Each zone cut from the strip is eluted separately with a small amount (about 2 ml) of water, and the eluate is analyzed for phosphodiesterase activity (5).

The extent of electro-osmosis may be determined by the extent of movement of the chloromycetin on the fifth strip. This compound may be located on the dry strip under an ultraviolet lamp.

TREATMENT OF DATA

Summarize your results and deliver a formal, oral report to the class.

QUESTIONS

1. From the direction of migration of the phosphodiesterase band, what can you conclude about the electrical charge upon this protein at pH 9.2? How must this conclusion be qualified by your observations of electro-osmosis?

2. Why is it advantageous to run the above electrophoresis at $4-10°$ rather than at room temperature?

3. How many other bands can you discern? What might some of them be?

References

1. Block, R. J., Durrum, E. L., and Zweig, G. 1955. Paper Chromatography and Paper Electrophoresis, A Manual. Academic Press. New York.
2. Corby, N. S., Kenner, G. W., and Todd, A. R. 1952. The Preparation of Tetra-esters of Pyrophosphoric Acid from Diesters of Phosphoric Acid by Means of Exchange Reactions. J. Chem. Soc., 1234–43.
3. MacDonald, H. J. 1955. Ionography, Electrophoresis in Stabilized Media, 187. Year Book Publishers. Chicago, Ill.
4. Butler, G. C. 1955. Phosphodiesterase from Snake Venom. In Methods in Enzymology, II, 561–5. S. P. Colowick and N. O. Kaplan, editors. Academic Press. New York.
5. Sinsheimer, R. L., and Koerner, J. F. 1952. A Purification of Venom Phosphodiesterase. J. Biol. Chem., 198, 293–6.
6. Williams, F. G., Jr., Pickels, E. G., and Durrum, E. L. 1955. Improved Hanging-Strip Paper Electrophoresis Technique. Science, 121, 829–30.

EXPERIMENT 13. ZONE ELECTROPHORESIS ON A STARCH SLAB (3 periods)

OBJECTIVE

This experiment will extend the technique of zone electrophoresis to a medium other than paper and to a technique which may be used for separations on a preparative scale.

PRINCIPAL EQUIPMENT AND SUPPLIES

Adjustable power source; $0-500$ volts and $0-200$ ma
Plastic trough for the starch slab (1)
Washed potato starch (1)
Snake venom, Crotalus adamantus*
Bis(p-nitrophenol)phosphate*
Barbiturate buffer; 0.05 M, pH 9.2
Protein dye and fixative*

PROCEDURE

Assemble the apparatus as described in ref. (1), and fill the trough with a slurry of starch in barbiturate buffer of pH 9.2. Dissolve 2 mg of the dried venom in 2 ml of buffer; analyze the solution for protein by the method given in Appendix II and for phosphodiesterase by the method used in Experiment 12. Add 1.5 ml of the venom solution to the center of the starch slab and start the electrophoresis at a potential of 200 v. At periods of $2-3$ hr, disconnect the power supply, open the trough compartment, and press a strip of filter paper gently but uniformly against the top of the slab. The paper strip should pick up protein wherever it occurs on the starch, and after the strip is dyed with bromphenol blue by the procedure employed in Experiment 12, it will serve as a guide to the progress of the electrophoresis. Make any changes in the voltage that are indicated from these guide strips and continue the electrophoresis until good separation of the bands is indicated.

At the end of the electrophoresis, the power supply is disconnected and a final guide strip is stained for protein. The starch slab is sliced into segments and eluted with water (1). The eluate from each segment is analyzed for protein and phosphodiesterase as above.

The fractions that have phosphodiesterase activity should be recovered as the dry (lyophilized) powder, and the percentage of recovery of this enzyme and the change in specific activity should be calculated.

TREATMENT OF DATA

Summarize your results and deliver a formal, oral report to the class.

---

*See Experiment 12

QUESTIONS

1. What advantage does starch have for electrophoresis of proteins as compared with other common media such as paper or silica gel? (See ref. [2].)

2. How large an amount of protein could be separated at one time in the apparatus you used?

References

1. Paigen, K. 1956. Convenient Starch Electrophoresis Apparatus. Anal. Chem., 28, 284–6.

2. Kunkel, H. G. 1954. Zone Electrophoresis. In Methods of Biochemical Analysis, I, 141–71. D. Glick, editor. Interscience Publishers. New York.

BIOCHEMISTRY OF ENZYMES

Introduction

The experiments in this section are intended to provide experience with a diversity of techniques, assays, and materials commonly met in studies involving enzymes. A variety of subjects of study, related only in that enzymes are involved as catalysts or as chemical substances, are now classified under the title Enzymology, and a correspondingly great number of techniques have been applied to these problems. Selection of appropriate experiments poses a real problem. Unfortunately, also, some of the important methods in present-day Enzymology are not readily adapted to a laboratory course. These include studies of the mechanism of enzyme action or biosynthetic schemes which require as reagents additional enzymes that are not readily available. With this limitation in mind, experiments have been selected which will supply a foundation of experience to help in application of major methods to later investigations.

The following points are of particular importance in working with enzymes.

1. Enzymes should be treated gently and kept cold, because they frequently lose activity upon exposure to heat, organic solvents, surface denaturation (foaming), or extremes of pH.

2. Many enzymes are inactivated by traces of contaminants such as heavy metals. Use clean equipment and distilled water, and avoid metal

spatulas and stirrers. Cysteine $(0.001 \underline{M})$, which reacts with heavy metal ions and also acts as a reducing agent, or Versene $(0.001 \underline{M})$, which reacts with metal ions, may prevent inactivation and may also restore activity.

3. Avoid microbial contamination of your preparations. Materials to be kept for several days should be preserved by freezing, addition of 0.1% merthiolate or a layer of toluene, because molds can grow even at 0° on moist biological material. Ascertain, first, that enzyme activity is not destroyed by the preservative treatment.

4. It is of prime importance that enzyme assays be run with adequate controls. Enzyme reactions are often influenced by variables other than those under study; lack of knowledge of these variables will lead to incorrect conclusions as to the meaning of the data. Blanks and controls provide a measure of the extent to which results are influenced by unknown factors. A blank may be defined as a sample in which no reaction has occurred. For example, a blank could contain all ingredients of the assay mixture except the active enzyme which is replaced by heated, inactive enzyme. This blank would measure the extent of non-enzymatic reactions. Other blanks might lack substrate or enzyme, or might contain all components with some provision for termination of the reaction at the moment of mixing. A control could be a complete reaction mixture, but with a known amount of purified or standardized enzyme acting under standard conditions. Each blank or control provides different information. Controls and blanks are not always described in the following experiments because if you must design adequate controls for yourself you will develop a more thoughtful attitude regarding your results.

5. Other important points (often overlooked by students) in the performance of assays are to prepare duplicate or triplicate samples to obtain more reliable results; to perform operations in the identical manner each time; to mix solutions well after addition of each reagent; to work with a care proportionate to the expected degree of accuracy (1% to 5% in enzyme work); and to carry out the reaction at a fixed, controlled temperature. Assays are generally designed so that all components except the one to be determined are in excess; therefore, extreme accuracy in pipetting is necessary only with the solution of the material to be determined.

6. Students are often too optimistic regarding the purity of chemicals. It is probably safe to assume that the name on the label represents at least half of the material in the bottle, although even this has not been so in some cases. It is wise to test for impurities and when they are found to purify commercial organic chemicals (by crystallization, sublimation, distillation, chromatographic techniques, etc.). Impurities may interfere by acting as inhibitors or substrates of enzymes, or interfere with the assay analysis, for example.

It is intended that Experiments 14 to 18 of this section be done first. The other experiments can be performed in any order, although the sequence given is a satisfactory one. Periods represent the time required by a pair of students, provided all reagents, etc., are available.

Certain books are of value for many of the experiments in this section, and for work with enzymes in general. Manometric Techniques (1) describes numerous techniques. Methods in Enzymology (2) is a compendium which contains most of the techniques used in the following experiments. The Enzymes (3) describes many of the properties of the known enzymes. Advances in Enzymology (4) yearly summarizes knowledge in selected areas. Biochemical Preparations (5) describes well-tested methods for the preparation of compounds of biological interest. Students will find it well worth while to become acquainted with these books. Outlines of Enzyme Chemistry (6) provides a good introduction to the subject.

QUESTIONS

1. What information may be gained from each of the blanks and controls listed above?

2. Name six Journals that frequently contain articles dealing with enzymes.

References

1. Umbreit, W. W., Burris, R. H., and Stauffer, J. F. 1957. Manometric Techniques, 3rd ed. Burgess Publishing Co. Minneapolis, Minn.
2. Methods in Enzymology (Vols. I through IV). S. P. Colowick and N. O. Kaplan, editors. 1955–57. Academic Press. New York.
3. The Enzymes (Vols. I through IV). J. B. Sumner and K. Myrback, editors. 1950. Academic Press. New York.
4. Advances in Enzymology (yearly). F. F. Nord, editor. Interscience Publishers. New York.
5. Biochemical Preparations (published at intervals). John Wiley and Sons. New York.
6. Neilands, J. B., and Stumpf, P. K. 1955. Outlines of Enzyme Chemistry. John Wiley and Sons. New York.

EXPERIMENT 14. SOME PROPERTIES OF PURIFIED ENZYMES
AS ILLUSTRATED WITH CHYMOTRYPSIN (1 period)

OBJECTIVES

This experiment is intended to provide an introduction to the experimental study of enzymes by a direct demonstration of enzyme action. You will observe the effects of several variables upon enzyme activity.

PRINCIPLES

Enzyme activities are measured by the rates at which they cause physical or chemical changes. For example, chymotrypsin can be assayed by its ability to clot milk; the time required for clotting varies in approximately inverse relation to the amount of enzyme present (1, 2).

The extent to which an enzyme-catalyzed reaction proceeds depends on concentration of enzyme, concentration of substrate, presence of inhibitors and cofactors, chemical structure of substrate, temperature, pH, and oxidation-reduction conditions (3). Some enzymes can exist in inactive forms; for example, chymotrypsin is the trypsin- or chymotrypsin-activated form of chymotrypsinogen.

PRINCIPAL EQUIPMENT AND SUPPLIES

Water bath at 37°
Stop watch
Buffer: Mix 120 ml 0.1 $\underline{M}$ tris(hydroxymethyl)aminomethane (1.45 g/120 ml) with 160 ml 0.05 $\underline{M}$ cacodylic acid (1.1 g/160 ml) and adjust to pH 7.5
Chymotrypsinogen (crystalline): 10 mg in 6 ml buffer
Trypsin: 1 mg in 5 ml buffer
Skim milk solution: 1 g dry skim milk (Instant Starlac works well) in 100 ml buffer plus 1 ml 3 $\underline{M}$ $CaCl_2$

PROCEDURE

1. *Action of Trypsin on Chymotrypsinogen*

In the first part of this experiment, the rate at which trypsin converts chymotrypsinogen to chymotrypsin will be determined.

Pipet 5 ml of buffer into each of 9 numbered test tubes and pipet 5 ml of milk solution into each of 10 other numbered test tubes. Put the latter tubes into the water bath at 37°.

Mix 2.7 ml of chymotrypsinogen and 0.3 ml of trypsin solution. Put this tube into the bath immediately and start the stop watch. At 1, 5, 10, 20, 30, and 40 min, pipet 0.10 ml of this mixture into 5 ml of buffer in a tube, mix, and immediately add 0.20 ml of the diluted mixture to 5 ml of milk solution. Mix the contents and incubate at 37° to determine chymotrypsin activity. For each sample, record the time of addition of enzyme to milk and the time when curds first appear (look at the meniscus).

Use the four remaining tubes of milk solution as blanks to test the action of similarly diluted trypsin, chymotrypsinogen, buffer alone, and a sample of activated enzyme that has been heated to boiling and cooled.

2. Clotting Time of Milk as a Function of Chymotrypsin Concentration and of Temperature

Add 0.20 ml of various dilutions (1 : 10 to 1 : 500) of your activated chymotrypsinogen, prepared above, to 5 ml of milk solution and determine clotting time as above. Also determine the clotting time of milk by activated enzyme (diluted 1 : 10, 1 : 30, and 1 : 100) at room temperature.

Store your activated enzyme in the refrigerator (for use in the next experiment).

TREATMENT OF DATA

Record this experiment and subsequent ones in systematic form in your notebook. Organize the data into graphs, and, in the second part of the experiment, include a plot of the clotting time against the reciprocal of enzyme concentration, and then plot the amount of chymotrypsin formed vs. time. Consider particularly the relation between clotting time and enzyme concentration.

QUESTIONS

1. What is the chemical action of trypsin on chymotrypsinogen?
2. What is involved in the clotting of milk? What industrial importance has this process?
3. What is the apparent function of chymotrypsin in vivo?
4. How can the molecular weight of an enzyme be determined? What is the molecular weight of chymotrypsin?
5. What is the molar concentration of chymotrypsin in your assay mixture for the first part of this experiment?

References

1. Gorini, L., and Lanzavecchia, G. 1954. Recherches sur le mechanisme de production d'une proteinase bacterienne. Biochim. et Biophys. Acta, 14, 407–14.
2. Berridge, N. J. 1954. Rennin and the Clotting of Milk. Advances in Enzymology, 15, 423–48. Interscience Publishers. New York.
3. Sumner, J. B., and Myrbäck, K. 1950. Introduction. The Enzymes, I, 1–27. Academic Press. New York.

EXPERIMENT 15. ENZYME ASSAYS WITH THE COLORIMETER
(1 period)

OBJECTIVE

This experiment is intended to acquaint you with colorimetric methods of enzyme assay.

PRINCIPLES

Enzymes are assayed by the extent of their action rather than by measurement of the absolute amounts of enzyme protein. Thus, when we speak of the presence of an enzyme in a preparation we actually mean that a chemical reaction (presumably catalyzed by the enzyme) has been measured, rather than that the enzyme protein has been identified. This way of considering enzymes is quite adequate, because the assumption that reactions in biological systems are catalyzed by enzymes is generally correct. In each case, however, development of a reliable enzyme assay in which the measured reaction will actually be proportional to the amount of enzyme present is vital.

All enzyme assays depend on the same principle: determination (by physical, chemical, or biological means) of the amount of a reactant that disappears or of a product that is formed as a function of time. One of three general procedures may be applied. The first, used in Experiment 14, involves mixing the enzyme and substrate in an appropriate reaction mixture and observing the time required for some definite change to occur. A second type of assay involves mixing the enzyme and substrate and measuring, periodically or continuously, some continuous change such as increased light absorption or decreased gas pressure. The third method is that used in the present experiment: enzyme and substrate are mixed and allowed to react for a fixed time; then the reaction is stopped, and a quantitative determination is made of some reactant or product. Trichloracetic acid (TCA) in a 5% concentration is often used to stop enzyme reactions since it is an effective enzyme inactivator and precipitant. Another protein precipitant is 3% $HClO_4$ which, unlike TCA, does not absorb ultraviolet light and can be removed by neutralization with KOH and precipitation of $KClO_4$ in the cold. However, $HClO_4$ will hydrolyze some acid-labile compounds. Enzyme reactions may be stopped also by boiling, adding alkali, alcohol, etc. After the reaction is halted, the material to be assayed is usually determined without being separated extensively from other compounds in the reaction mixture.

The selection of an assay procedure depends both on the purpose for which it is to be used and on its reproducibility, sensitivity, simplicity, and precision. For example, simplicity is more important that great precision in an assay to be used during enzyme purification, although for kinetic studies the opposite is true. Many types of assays have been described in the literature. Comparison of various methods for assay of one type of enzyme can provide some understanding of the problems involved in setting up a good assay; an excellent compilation of over a dozen methods for assay of proteolytic enzymes (1) is available. When you are using published methods, conformity to stated details of procedure is most important. Simplifications may seem obvious to you, but frequently a change in procedure will cause difficulties (e.g., high blank values or lack of proportionality between amount of reaction and concentration of enzyme). One exception to this generalization is that older published assays often can be made more sensitive by reducing all volumes in proportion.

Colorimetric methods are preferred for enzyme assays to any other means of quantitative determination in biochemistry because of their specificity, high sensitivity (0.1 to 100 μg of substrate), ease of performance, and the fact that many samples may be assayed at the same time.

The operation of photoelectric colorimeters (such as the Klett-Summerson instrument) varies somewhat from one instrument to another (2, 3). In general you first select light of a limited wavelength band, by means of a filter, diffraction grating, or prism, to provide maximal absorption by the desired material and minimal absorption by other substances, such as reagents, that are present. Second, you turn on the electricity and allow a period for the instrument to become stable. Third, you "zero" the instrument under two sets of conditions: with no light striking the photocell and then with the light passing through a blank solution, usually water, or a reagent blank (all reagents required to develop the color are added to water). Finally, you replace the blank with a sample and take a reading by balancing the photoelectric circuit and then reading the scale. Assays are most convenient if they obey Beer's Law, i.e., if colorimeter reading is proportional to amount of material (2, 3).

PRINCIPAL EQUIPMENT AND SUPPLIES

Water bath at 15° to 45°

Colorimeter and tubes

Activated chymotrypsin from previous experiment; students should share the same enzyme preparation to permit comparison of results

5 Mg/ml casein in 0.1 M borate buffer, pH 8.0 (1.24 g boric acid, 100 ml water, 3 ml 1 M KOH, plus 0.15 ml 3 M $CaCl_2$. Heat 15 min in a boiling-water bath to dissolve the protein and bring the volume to 200 ml with water)

Fresh solution of 5% trichloracetic acid (TCA)

Folin-Ciocalteu reagents (see Appendix II)

PROCEDURE

1. *Calibration of the Folin Method*

(Do this part while the samples of part 2 are stored for 1 hr.) Dilute a portion of your casein solution to 0.3 mg/ml with 0.5 N NaOH, and pipet 0.0, 0.1, 0.2, 0.3, 0.5, 0.7, and 1.0 ml aliquots into numbered test tubes. Add 0.5 N NaOH to make the total volume 1.0 ml in each tube, and run the Folin-Ciocalteu test for protein (see Appendix II). Plot color intensity vs. mg of protein.

2. *Proteolytic Activity of Chymotrypsin* (4)

Bring 20 ml of casein solution to a specific temperature between 15 and 45°. Put 3 ml of 5% TCA into each of 8 numbered test tubes. Mix 2 ml of casein solution with the TCA in the first tube. Add 0.20 ml of activated chymotrypsin to the remaining casein solution, mix by swirling or by inverting with parafilm over the mouth of the tube, and incubate at the specific temperature you have chosen. At 1, 2, 5, 10, 15, 20, and 30 min after addition of the enzyme, pipet 2 ml aliquots into the tubes containing TCA. One hr after completion of sampling, centrifuge all 8 tubes and run protein analyses in duplicate on 0.10 ml aliquots of each clear supernatant solution.

TREATMENT OF DATA

Plot your data and discuss the results; give reasons for lack of linearity of the plot. Compare the rate of reaction you obtained with the results found at other temperatures by other students.

QUESTIONS

1. What structures in a protein give color reactions with the Folin-Ciocalteu reagent and with the Biuret reagent? Would you expect all proteins to give the same amount of color per gram of protein?
2. Why is a red filter used for this assay?
3. Does Beer's Law hold for this assay?
4. List some requirements for a good assay method. How well do the milk-clotting assay and the one used in this experiment satisfy these requirements? Name, in order of merit, other possible assays for chymotrypsin.
5. How would you decide whether or not a crystalline solid isolated from biological material is an enzyme?
6. Compare the relative advantages in this assay of centrifugation, gravity filtration, and vacuum filtration.
7. What physical-chemical equation relates rate of a reaction and temperature?
8. Is there a commercial use for the peroxidase test? (See page 79.)
9. What role is peroxidase thought to have in vivo?
10. Why does heat destroy enzyme activity?

ALTERNATIVE EXPERIMENT, 15*

Peroxidase is an enzyme that catalyzes oxidation by H_2O_2 of phenols and aromatic amines (5). Guaiacol is oxidized to a complex colored product by H_2O_2 in the presence of peroxidase; the rate of color production can be used as a measure of enzyme activity.

Peroxidase, like most enzymes, can be inactivated by heating to boiling. However, if it is heated for 1 min or less, activity is partly recovered after a period of standing (6). The enzyme contains iron in a heme group, and various compounds that form complexes with iron, such as azide, cyanide, and hydroxylamine inhibit the enzyme activity.

PROCEDURE

Peel, wash, and grind a turnip in a blender (or chop it finely in a meat grinder). Squeeze the juice through a piece of cheesecloth. Add about 1 g of Hyflo Super Cel filter-aid and filter the juice. Dilute 1 ml of the juice to 200 ml with distilled water.

Place 0.1 millimole of liquid guaiacol (0.01 ml), together with 0.2 ml of 0.9% H_2O_2 and 4.7 ml of water in a test tube. Put 1.0 ml of diluted extract and 4 ml of water into a colorimeter tube. Pour the guaiacol-H_2O_2 solution into the colorimeter tube and start the stop watch. Pour the mixture quickly back into the empty tube and then into the colorimeter tube and immediately place it in the colorimeter (adjusted beforehand, with a blue filter against water). Record readings every 20 sec. Repeat the determination, using one-half and then twice as much extract. Also try the effect of extract which immediately before assay has been placed in a bath of boiling water for two minutes and then cooled. Test the effect of 0.01 M sodium azide (Caution: poison!) or 0.01 M hydroxylamine on the reaction. Alternatively, reversible heat inactivation (6) may be studied.

Make graphs of colorimeter readings against time, for each determination.

References

1. Davis, N. C., and Smith, E. L. 1955. Assay of Proteolytic Enzymes. In Methods of Biochemical Analysis, II, 215–57. D. Glick, editor. Interscience Publishers. New York.
2. Hiskey, C. F. 1955. Absorption Spectroscopy. In Physical Techniques in Biological Research, I, 73–130. G. Oster and A. W. Pollister, editors. Academic Press. New York.
3. West, W. 1949. Colorimetry, Photometric Analysis, Fluorimetry, and Turbidimetry. In Physical Methods of Organic Chemistry, 2nd ed., I, 1399–490. A. Weissberger, editor. Interscience Publishers. New York.

---

* This experiment was adapted from "Laboratory Experiments for Enzyme Chemistry" by Dr. A. K. Balls, Purdue University, Lafayette, Indiana.

4. Northrop, J. H., Kunitz, M., and Herriott, R. M.   1948. Crystalline Enzymes, 308–10. Columbia University Press. New York.
5. Theorell, H.   1951. In The Enzymes, II, 397–427. J. B. Sumner and K. Myrbäck, editors. Academic Press. New York.
6. Schwimmer, S.   1944.   Regeneration of Heat-Inactivated Peroxidase. J. Biol. Chem., 154, 487–95.

EXPERIMENT 16.   OPERATION OF THE pH METER AND DETERMINATION OF pK_a' (1 period)

OBJECTIVES

This experiment is intended to give you experience in operation of the pH meter, with the response of buffered and unbuffered solutions to acid and base, and with the determination of pK_a'.

PRINCIPLES

Since the activities of enzymes are dependent on the hydrogen ion activities, $[H^+]$, of their solutions it is important to adjust the pH of an enzyme assay mixture to a selected value, and also to be sure that there is no significant change during the course of the reaction (1). The determination of pH is most conveniently made with a pH meter (2, 3), although indicator dyes also can be used to estimate pH.

The pH of an enzyme assay mixture may be held constant during a reaction by the presence of a suitable buffer. A buffer is a mixture of an acid and its salt, such as acetic acid and sodium acetate, or more generally HA and A^-; or, what amounts to the same thing, a base and its salt, such as ammonia and ammonium chloride, or more generally B and HB^+. The pH of such a solution depends on the dissociation constant of the buffer and the ratio of acid to salt.

$$K_a' = \frac{[H^+](A^-)}{(HA)} \qquad [1]$$

Equation [1] may be converted into the convenient form of the Henderson-Hasselbalch equation (3):

$$pK_a' = pH + \log \frac{(HA)}{(A^-)} \qquad [2]$$

where $pK_a' = -\log K_a'$, by definition, and the prime mark as a superscript of the K_a denotes the fact that the value was determined and is valid only for a specified concentration. See ref. (3) for the effect of concentration on the value of pK_a'.

Equation [2] is particularly convenient for designing buffer mixtures for specified pH values. First, the maximum buffer capacity of any buffer occurs at the pH equal to its pK_a'. This is perhaps best shown in the familiar titration curve, pH vs. amount of alkali added (1), where it may be seen that the slope of the titration curve is minimal, i.e., the pH change is least per increment of alkali at the midpoint of the titration curve, and at this midpoint the pH equals pK_a'. Second, the buffer power becomes increasingly weak as the pH becomes increasingly lower or higher than the pK_a', until, at pH values about 1 pH unit above or below the pK_a', the buffer power becomes too weak for practical purposes. Third, the proper ratio of acid and anion for a buffer of required pH may be readily calculated by equation [2].

In addition to the relation of pK_a' to the pH range of adequate buffer action, the concentration of buffer determines its effectiveness or capacity. The concentration of buffer required in a given experiment depends on the amount of acid or base liberated by the reaction, but usually a concentration of 0.01 to 0.1 M buffer is present in enzyme assays. A table of buffers is given in Appendix IV. All compounds with the same pK_a' value are equally effective buffers if they are at equivalent concentrations, regardless of their structures. However, properties other than pK_a' must be considered in the choice of a buffer for work with enzymes, and these include toxicity, cost, ability to form complexes with metal ions, and interference with analytical methods.

A knowledge of the pK_a' of an acid is essential for intelligent use of the compound as a buffer, as has been indicated. The pK_a' may be calculated from titration data by several methods. First, the pH at the inflection point of a plot of pH vs. equivalents of base (or acid) added is equal to pK_a'. Second, pK_a' is equal to the pH at which the acid is half titrated (e.g., $(HA)/(A^-) = 1$). This point of half titration is readily found from the titration curve if the end point can be determined or calculated (by observation, from knowledge of the equivalents of acid added, or by assuming it to be 2 pH units above the approximate inflection point). Third, pK_a' can be calculated from any point on the titration curve by substitution of the experimental values for pH, (HA), and (A^-) in equation [2]. Several such calculations may be made for different points on the curve, and these values of pK_a' can be averaged. At extreme pH values (above 10 or below 3), acid or base required to titrate the solvent alone to each pH should be subtracted from the amount added to the solution.

PRINCIPAL EQUIPMENT AND SUPPLIES

pH meter
10 Ml burette graduated to read 0.05 ml
0.1 M NaH_2PO_4
0.1 M Na_2HPO_4
0.050 N NaOH
0.050 N HCl
0.05% Bromthymol blue
0.020 M uracil

PROCEDURE

1. *Buffer Action*

Read the directions for operation of the pH meter before you begin the experiment. Note particularly that the glass electrode is fragile! Calibrate the meter with a standard buffer, according to the directions, rinse the electrodes with distilled water, and blot them with absorbent tissue. Calculate the ratio of 0.1 M NaH_2PO_4 ($pK_a = 6.8$) and 0.1 M Na_2HPO_4 required to make a solution of pH 6.5; prepare 20 ml of this solution and add a few drops of bromthymol blue. Measure the pH of this solution and determine the amounts of acid and base required to lower and raise the pH one unit. Observe the changes of indicator color also. Repeat the experiment with 0.1 M phosphate, pH 5.5 and then with distilled water.

2. *Determination of* pK_a' *of Uracil*

Titrate 20 ml of 0.02 M uracil to pH 11.5 with 0.05 N NaOH and record the pH for every 0.50 ml of alkali added. Next titrate distilled water over the same range of pH, to obtain a correction for alkali required to change the pH of the solvent.

TREATMENT OF DATA

Plot your data for titration of uracil as pH vs. volume of NaOH added in excess of the amount required to bring water to the same pH; on the same graph show the fraction of uracil neutralized. Determine the pK_a' of uracil by all of the three methods mentioned under Principles.

QUESTIONS

1. Why is the pH of distilled water not necessarily 7.0?
2. In what ionized forms does alanine exist at pH 1.5, at pH 5.0, and at pH 11.0?
3. List two compounds useful as buffers at pH 5.0, two at pH 7.0, and two at pH 9.0. At what pH would uracil be the best buffer?
4. What method provides a primary standard to determine pH?
5. What are the principal parts of a pH meter (omit circuit diagrams)? What is the principle of operation?
6. What is the function of the temperature control knob on the pH meter?

References

1. Neilands, J. B., and Stumpf, P. K. 1955. Outlines of Enzyme Chemistry, Chap. 2. John Wiley and Sons. New York.
2. Willard, H. H., Merritt, L. L., Jr., and Dean, J. A. 1951. Instrumental Methods of Analysis, Chap. 13. D. Van Nostrand Co. New York.
3. Bates, R. G. 1954. Electrometric pH Determinations. John Wiley and Sons. New York.

EXPERIMENT 17. DETERMINATION OF ESTERASE ACTIVITY
BY TITRATION * (1 period)

OBJECTIVES

This experiment provides experience in the estimation of enzyme activity by titration, and in preparation of a crude enzyme from a plant source.

PRINCIPLES

Numerous biochemical reactions result in the uptake or release of acid; therefore, determination of the rate at which these reactions proceed is possible by titration of the liberated alkali or acid. This titration is done in such a way as to keep the pH within a narrow range to avoid affecting the enzyme activity. Instruments called "pH-stats" have been designed which automatically add alkali to keep the pH constant, and which record the volume added as a function of time (1). These instruments can also plot titration curves. In the present experiment the activity of an esterase on triacetin will be measured by titrating manually the acid released as a function of time.

PRINCIPAL EQUIPMENT AND SUPPLIES

Blender (optional)
pH meter
Stop watch
10 Ml burette graduated to read 0.05 ml
0.050 N NaOH
10 G of pumpkin seeds (or 1 orange)
2% triacetin in water
Ethyl acetate
Tributyrin
Lyophil apparatus (optional)
Versene

PROCEDURE

1. *Preparation of the Crude Enzyme*

Remove the shells from 10 g of pumpkin seeds and grind the kernels to a paste with a few ml of water. Gradually stir in more water until 40 ml have been added. Alternatively, disrupt the seeds in a blender with 40 ml 0.002% Versene. Centrifuge at 3000 rpm for 5 min, remove the top layer of fat as well as possible, and pour off the cloudy supernatant fluid. Use this supernatant fluid as a crude preparation of acetyl ester-

---

* This experiment was derived from the syllabus of Laboratory Experiments for Enzyme Chemistry by Dr. A. K. Balls, Purdue University, Lafayette, Indiana.

ase. (Store the preparation in the cold; toluene is not satisfactory as a preservative in this instance.)

Optional: lyophillize 10 to 20 ml of this supernatant solution (2, 3). When dry, store the product in the cold. The fat may be removed by washing the lyophillized powder 2 or 3 times with a little petroleum ether and allowing the residue to dry in the air.

Fluid squeezed from the grated outer part of an orange may be used as a source of esterase, in place of pumpkin seeds (1, 4).

2. *Estimation of Acetyl Esterase*

Place 50 ml of 2% triacetin and 0.1 ml of saturated KCl solution in a 150 ml beaker; insert the electrodes from the pH meter and a small stirrer. Fill the burette with 0.05 $\underline{N}$ NaOH and place it above the beaker. Adjust the solution to slightly above pH 6.5. Add 3 to 4 ml of esterase solution, start the stirrer, readjust the pH if necessary, and read the burette. Start the stop watch when the pH drops to 6.5. When the galvanometer needle of the pH·meter has moved a few divisions (or 0.1 pH unit) to the acid side of pH 6.5, bring the solution to slightly above pH 6.5 with alkali, and record the burette reading. When the pH returns to 6.5 again record the time. Repeat this process of alternate pH adjustment and time measurement at least four more times.

Repeat the experiment with enzyme that has been placed in a boiling water bath for 5 min and then cooled; this process will serve as a control to correct for spontaneous hydrolysis of the triacetin and for CO_2 uptake from the air.

If time permits, repeat the assay with another substrate, such as ethyl acetate or tributyrin.

TREATMENT OF DATA

Plot milli-equivalents (meq) of base added against the time recorded at each return to pH 6.5. Calculate the rate of enzymatic hydrolysis of triacetin in meq/min from the slope of your graph. Write an equation for the reaction (5).

QUESTIONS

1. Are many plant enzymes known? List six of them.

2. Why are enzymes dried by lyophillization rather than by evaporation of the solvent at room temperature?

3. What other assay methods could be used to estimate this enzyme?

4. What results would you obtain with this assay if the seeds were extracted with 0.1 $\underline{M}$ phosphate buffer, pH 6.5?

5. Under what conditions could peptidase activity be measured by this titration method?

6. Why was KCl added in this experiment? How would you find out if your explanation is correct?

7. How could you avoid interference by atmospheric CO_2?

References

1. Neilands, J. B., and Cannon, M. D. 1955. Automatic Recording pH Instrumenta-
   tion. Anal. Chem., 27, 29–33.
2. Kabat, E. A., and Mayer, M. M. 1948. Experimental Immunochemistry, 437–40.
   Charles C. Thomas. Springfield, Ill.
3. Campbell, D. H., and Pressman, D. 1944. A Simple Lyophil Apparatus. Science,
   99, 285–6.
4. Jansen, E. F., Jang, R., and MacDonnell, L. R. 1947. Citrus Acetylesterase.
   Arch. Biochem., 15, 415–31.
5. Mattson, F. H., and Beck, L. W. 1956. The Specificity of Pancreatic Lipase for
   the Primary Hydroxyl Groups of Glycerides. J. Biol. Chem., 219, 735–40.

EXPERIMENT 18. THE SPECTROPHOTOMETER (1 period)

OBJECTIVES

The purpose of this experiment is to increase your familiarity with the operation of the spectrophotometer. The instrument will be used to determine the pK_a' of an ionizable, light-absorbing compound (uracil), from measurements of absorption spectra at 3 pH values.

PRINCIPLES

The spectrophotometer is an instrument that measures the fraction of incident light of a given wavelength transmitted by a solution. Readings are commonly recorded as optical density, equal to $\log_{10}$ (incident/transmitted light intensity), since this quantity is often proportional to the concentration of the light-absorbing compound (Beer's Law). Both the spectrophotometer and the colorimeter measure optical densities; however, the former has great advantages of accuracy, sensitivity, broad spectral range, and sharp selection of wavelength. See an article on spectrophotometry and colorimetry (1, 2) for a review of this subject.

The spectrophotometer is used principally for three different types of determinations in Enzymology. First, the concentration of a compound can be determined by measurement of optical density at one wavelength, under conditions where no changes occur, provided that the extinction coefficient of the compound is known. Second, the course of a reaction can be observed in a complete assay system by measurement of the rate of production or disappearance of a light-absorbing compound (see Experiments 19, 20, and 29). Commercially available recording spectrophotometers that make a continuous plot of optical density vs. time are most useful for these measurements. A third sort of measurement, valuable for identification of compounds, requires the determination of optical density as a function of wavelength (1). Here again, instruments are available which trace such plots automatically.

PRINCIPAL EQUIPMENT AND SUPPLIES

Beckman model DU spectrophotometer and 4 quartz cuvettes
0.02 $\underline{M}$ uracil
0.05 $\underline{M}$ glycine buffer, pH 9.5
0.01 $\underline{N}$ NaOH
0.01 $\underline{N}$ HCl

PROCEDURE

1. *Operation of the Spectrophotometer*

Familiarize yourself with the spectrophotometer by means of the instructions that come with the instrument. Note that Corex and silica cuvettes are expensive: take care of them accordingly! Wipe optical surfaces only

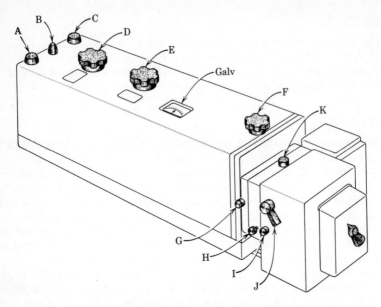

Figure 14. Beckman spectrophotometer.

with lens paper; after use, rinse cuvettes thoroughly and allow them to drain dry.

The essential steps in operation of the Beckman spectrophotometer are listed below (see Figure 14):

a. Attach the proper light source to the instrument: hydrogen lamp for ultraviolet (U.V.) below 350 mμ, and tungsten at longer wavelengths. The hydrogen lamp is turned on by setting the right-hand dial of the power source to a 2-o'clock position and turning on the left-hand switch. After 1 min the button of the power source is pressed and the right-hand dial is turned back to a 10-o'clock position to reduce the temperature of the filaments for longer tube life.

b. Turn left center switch (B) of the spectrophotometer to "check." Allow the instrument to become steady.

c. Select the desired wavelength with dial (D), top center.

d. Select the proper phototube with knob (I), extreme right front (in above 625 mμ, out below 625 mμ).

e. Filter knob (G) is out between 320 and 400 mμ, and is in for all other wavelengths.

f. Place a cuvette, silica below 320 mμ, Corex at longer wavelengths, containing the blank (distilled water) into the first compartment of the cuvette holder and insert cuvettes containing samples into the other three compartments. Set the loaded holder squarely into the tray inside the cuvette compartment (K) and cover the compartment.

g. With the shutter switch (J) off, adjust the dark current knob (A), top left front, so that the galvanometer (Galv) reads "0."

h. Turn the sensitivity dial (C), top left rear, to about three revolutions from the counterclockwise limit. Place the blank in the light path by means of knob (H), open the shutter (J), and adjust the galvanometer to "0" using slit dial (F) top right, and dial (C) for coarse and fine adjustments, respectively.

i. With the instrument now adjusted for shutter both closed (dial A) and open (dials C and F), move a sample into the light path by means of knob (H). Turn switch (B) from "check" to "1" and with shutter (J) open adjust the galvanometer to "0" with the Transmittance/Absorbance dial (E). Record the reading of the optical density on the lower scale next to knob (E).

j. Repeat the measurements with the other samples. If the optical density is greater than 1.0, turn switch (B) to "0.1." This increases the response of the instrument tenfold. Read optical density as usual, but add 1.00 to your reading.

k. When you are finished, clean the cuvettes and turn off all 3 switches (B, J, and light).

2. *Color-Wavelength Relations*

Place a white card in the sample well (K). Turn the tungsten light on, set the slit at 2.0, and darken the room if possible. Slowly rotate the wavelength knob and record the colors observed on the card as a function of wavelength. What do you observe at 570 mμ when you decrease the slit width?

3. *Spectra of Uracil at Various pH Values* (3)

Prepare 10^{-4} M solutions of uracil in 0.01 N HCl, 0.01 N NaOH, and 0.05 M glycine buffer of pH 9.5. Accurately determine the pH of the buffered solution. Measure optical densities of the three uracil solutions and of the three solvents (in the absence of uracil) at 10 mμ intervals from 310 to 220 mμ.

TREATMENT OF DATA

Prepare the following:

1. A table of color vs. wavelength.
2. A graph of your data for the optical density of uracil (corrected for any absorption by solvent) against wavelength
3. A calculation of the pK_a' of uracil (4).

Assume the spectrum in acid is that of unionized uracil, in alkali that of totally ionized uracil, and the spectrum of uracil in buffer results from absorption by both these forms and only these forms. Calculate the degree of ionization in the buffered solution from data at a wavelength where the two forms have very different optical densities. Check this result by repeating the calculations with data obtained at another wavelength. From these results and the measured pH calculate the pK_a' of uracil. Compare your results with those obtained in Experiment 16.

QUESTIONS

1. Define optical density, transmittance, and molar extinction coefficient. From your data, calculate the molar extinction coefficient of uracil at 260 mμ and pH 7.0.

2. Sketch the light-path of the spectrophotometer. What is the function of each mechanical and optical part of the instrument?

3. How can you distinguish a silica cuvette from a glass one?

4. Do your data display any isosbestic points for uracil?

5. Compare the merits of determination of pK_a' with the spectrophotometer and with the pH meter, as dependent on properties of the compound under study.

6. Protein is sometimes estimated by an optical density measurement at its absorption maximum (280 mμ). Biological materials often contain nucleic acids which absorb strongly at 280 mμ and even more strongly at 260 mμ. How and under what conditions can you estimate protein in the presence of nucleic acids, from light absorption data?

References

1. Scott, J. F. 1955. Ultraviolet Absorption Spectroscopy. In Physical Techniques in Biological Research, I, 131–203. G. Oster and A. W. Pollister, editors. Academic Press. New York.
2. West, W. 1949. Spectroscopy and Spectrophotometry. In Physical Methods of Organic Chemistry, 2nd ed., I, 1242–398. A. Weissberger, editor. Interscience Publishers. New York.
3. Shugar, D., and Fox, J. J. 1952. Spectrophotometric Studies of Nucleic Acid Derivatives and Related Compounds as a Function of pH. Biochim. et Biophys. Acta, 9, 199–218.
4. Flexser, L. A., Hammett, L. P., and Dingwall, A. 1935. The Determination of Ionization by Ultraviolet Spectrophotometry; Its Validity and Its Application to the Measurement of the Strength of Very Weak Bases. J. Am. Chem. Soc., 57, 2103–115.

EXPERIMENT 19. PURIFICATION OF AN ENZYME: YEAST ALCOHOL DEHYDROGENASE (4 periods)

OBJECTIVES

This experiment is intended to provide experience with several techniques commonly used in the purification of enzymes.

PRINCIPLES

Purified enzymes are essential for the study of reaction mechanisms (kinetics, nature of attached groups, cofactor requirements, etc.) and are desirable for application as analytical or preparative reagents. Some of the techniques used to obtain purified enzymes are different from those used in other areas of biochemistry, mainly because of the large amount of starting material often required, the lability of enzymes, and their low concentration in natural materials. The methods most frequently employed are selective denaturation of proteins by heat or acid, fractional precipitation with ammonium sulfate, low temperature fractionation with alcohol or acetone, selective adsorption on and elution from calcium phosphate or alumina gel, and crystallization (1, 2). Zone electrophoresis on starch (see Experiment 13) is a newer method of considerable promise, as is chromatography on cellulose columns (3, 4). For the original purification of an enzyme, one must try various methods and select those that result in the best balance of purification and recovery. Procedures which result in good increases of purity often also give low yields. If an established method is to be used, it is important to follow the given directions exactly. Enzyme activity and protein content should be determined at each major step, and these should parallel published results.

Some comments should be made on frequently used techniques.

1. The results often depend on the brand, strain, or variety of starting material; obtain the same type as was used previously, if possible.

2. When saturated $(NH_4)_2SO_4$ is used as a precipitating agent it is important to observe the temperature at which saturation was achieved, since solubility of the salt depends markedly on temperature. The solubilities of enzymes also vary with temperature. In some cases, enzymes are less soluble in $(NH_4)_2SO_4$ solutions at higher temperatures than at lower.

3. The use of organic solvents as enzyme precipitants is likely to result in denaturation unless local excesses of solvent are avoided and the entire system is kept cold. The precooled solvent should be dripped into the stirred, cold aqueous solution as rapidly as possible while keeping the temperature at the desired value. Dry ice and water plus Cellosolve, alcohol, or acetone baths are used to keep the temperature low (sometimes as cold as $-20°$) and the process is carried out in a cold room. The desired temperature may be held constant by choice of a solvent mixture with the appropriate freezing point. Most steps in enzyme purification are best carried out at low temperatures $(0°)$.

4. Precipitate may be removed from supernatant fluid by three means: centrifugation (5), gravity filtration (6), or vacuum filtration (6).

a. Centrifugation is frequently applied for separation of protein precipitates and is especially useful for removing gummy or gelatinous materials. One defect of the method is that the pellet may contain up to 90% mother liquor so that a clean separation is not achieved in the precipitate fraction. Also, centrifugation of large volumes is quite tedious. Sometimes the precipitate will settle after standing, and then much of the supernatant fluid can be removed by decantation.

b. Gravity filtration through soft, thick, fluted papers is often very convenient for handling large volumes, especially if the filters are kept filled from a reservoir. (To maintain a constant level in the filter, provide the reservoir bottle with a two-hole stopper through which pass a siphon to the filter and a tube for entry of air reaching to the bottom; set the bottle above the filter so that the air inlet is 1 cm below the desired height of the liquid in the filter. See the reservoir in Figure 5.)

c. Filtration under vacuum through a Buchner funnel can be rapid and convenient; however, precipitates of biological material often clog the filter. Filter aids such as Celite or shredded filter-paper pulp, put on the filter as a pad and mixed with the suspension, may prevent clogging. Air should not be pulled through the precipitate because it may denature the enzyme. Removal of very coarse material such as ground tissue may be accomplished by rapid filtration through a pellet of glass wool, or by squeezing the liquid through cheesecloth before filtration.

5. Remember that precipitates contain considerable quantities of mother liquor. The amount may be sufficient to upset subsequent procedures by introduction of salts, solvents, etc. Dialysis (7) is used to remove salts (and also to precipitate some proteins), but dialysis of concentrated salt solutions can result in pH changes which may damage enzymes. Dialysis against dilute buffers or salt solutions is helpful in these cases. The precipitates will also contain denatured, insoluble proteins, and attempts to dissolve a precipitate completely will result in too great dilution of the enzyme.

6. Enzyme inactivation is most likely in the later stages of purification, when the protective action of other proteins is absent. To avoid loss of activity, use clean glassware, prevent contamination by heavy metals (from tap water, for example), avoid foam formation, and keep the enzyme solution concentrated. Addition of a chelating agent such as Versene (ethylenediamine tetra acetic acid), a reducing agent like cysteine (0.001 $\underline{M}$), or even a substrate may protect the enzyme.

7. Crystallization of an enzyme, if successful at all, is liable to be a slow process; set concentrated solutions in the cold for several days. Seeding with crystals of the desired material and scratching the glass or warming and cooling slightly may help (8).

PRINCIPAL EQUIPMENT AND SUPPLIES

Mechanical stirrer
High-speed refrigerated centrifuge
Ball mill
100 G of dry bakers' yeast

600 Ml 0.066 $\underline{M}$ Na$_2$HPO$_4$

500 Ml of cold acetone

Four 250-ml glass centrifuge bottles

Dialysis tubing

10 Liters of 0.001 $\underline{M}$ phosphate buffer, pH 7.5 (cold)

10 Ml of 3 $\underline{M}$ ethanol (2 ml 95% ethanol + 8 ml H$_2$O)

50 Ml of 0.06 $\underline{M}$ sodium pyrophosphate, pH 8.5 (mol wt Na$_4$P$_2$O$_7$·10H$_2$O = 446.1)

5 Ml 0.015 $\underline{M}$ enzymatically assayed DPN (mol wt DPN·4H$_2$O = 735)*

100 Ml 0.1% bovine serum albumin in 0.01 $\underline{M}$ potassium phosphate, pH 7.5

PROCEDURE

Alcohol dehydrogenase catalyzes the oxidation of ethanol by DPN (co-enzyme I). The reaction is conveniently assayed by measuring the rate of appearance of reduced DPN (DPNH) by means of the change in optical density at 340 mμ.

Prepare crystalline alcohol dehydrogenase from bakers' yeast according to the procedure of Racker (9). Determine the amounts of enzyme and protein at various stages as you proceed, at steps noted below. Do not discard any fraction until you are certain that it contains relatively little dehydrogenase activity. This preparation is not easy, especially since there may be considerable periods of storage. However, much can be learned even if the final crystalline product is not obtained.

1. *Purification Procedure*†

Fleischmann's bakers' yeast is obtained in 1 lb cakes, crumbled, and dried between two sheets of paper in a thin layer for 4 to 5 days at room temperature; or Fleischmann's 20-40 dry yeast can be used.‡ The dry yeast is finely ground in a ball mill at 0° for 16 hr and stored in the cold room in a well-stoppered container. 100 g of the dry yeast powder is extracted with 300 ml of 0.066 $\underline{M}$ disodium phosphate for 2 hr at 37° with continuous stirring, followed by extraction at room temperature for an additional 3 hr. The yeast residue is then removed by centrifugation at 13,000 rpm for 15 min (assay). The supernatant solution is quickly brought to 55° and maintained at this temperature in a water bath for 15 min; after cooling, the mixture is centrifuged. At this stage the clear supernatant fluid can be stored at 0° overnight (assay). Further steps are carried out in a cold room.

To each 100 ml of the yeast extract at $-2°$, 50 ml of acetone at $-10°$ is added over a period of about 20 min, slowly at first, so as to keep the enzyme at $-2°$, in an ice-salt bath $(-6°)$. Do not allow freezing to occur.

---

* This concentration of DPN gives more consistent results than does 0.0015 $\underline{M}$ DPN (E. Racker, personal communication).

† Portions of this procedure are quoted from ref. (9) with permission of the publishers and author.

‡ Standard Brands, Inc., 595 Madison Ave., New York, N. Y.

The resulting precipitate is separated by centrifugation in glass bottles for 10 min at 2500 rpm at 0°. To the supernatant solution (assay), 55 ml of acetone ($-10°$) is added for each 100 ml of the yeast extract, the temperature being kept at $-2°$. The mixture is centrifuged at 0°, and the supernatant solution discarded. The precipitate is suspended in about 50 ml of cold 0.01% Versene and 0.001 M cysteine and dialyzed for 3 hr against 3 3-liter batches of 0.001 M phosphate buffer, pH 7.5 (assay).* The precipitate is removed by centrifugation. To the clear supernatant solution 36 g of finely ground solid ammonium sulfate is added per 100 ml of solution. After standing at 0° for 30 min, the mixture is centrifuged at 15,000 rpm for 20 min at 0°. Supernatant and precipitate (assay)† may be stored at $-20°$. The precipitate is dissolved in 20 ml of distilled water (assay), and 4 g of ammonium sulfate is added. The precipitate is centrifuged off (assay). From the supernatant solution, the alcohol dehydrogenase enzyme starts to crystallize a few minutes after the further addition of 4 g of ammonium sulfate, which is added in small portions during the course of several hours. The crystals are collected by centrifugation and resuspended in a small volume of distilled water (assay), from which they start to recrystallize almost immediately on addition of small volumes of saturated ammonium sulfate solution (assay). A typical protocol showing data on the specific activity of the various fractions and the yields obtained is presented in Table XII (9). Purification should be completed in the third period.

"The enzyme is most conveniently stored in 50% saturated ammonium sulfate at $-20°$.

---

* The enzyme is more stable at ionic strengths greater than 0.1 and in the presence of 0.001 M cysteine (A. L. Marr, personal communication) or 0.01% Versene (F. A. Loewus, personal communication). For dialysis, a bag is made by tightly knotting the end of a piece of cellulose sausage casing. The bag is first soaked in water (dilute solutions of reducing agents have been found useful for pretreatment in some cases) and tested for leaks by filling it with water and applying pressure. The enzyme solution is poured in, the bag is tied off at the top and is suspended in a large beaker of water (or salt solution, etc., in other cases). For rapid dialysis, it is best to stir the solutions both inside and outside the bag. This is accomplished readily by hanging the bag from the horizontal arm of an L-shaped rod; the rod is caused to rotate slowly by inserting its vertical arm into the chuck of a motor. The solution outside the bag can be changed continually or intermittently.

† With some batches of yeast, these fractionations were found to place considerable amounts of alcohol dehydrogenase in the wrong fractions (F. A. Loewus, personal communication). Further fractionation may be required.

TABLE XII. PURIFICATION OF ALCOHOL DEHYDROGENASE

FROM BAKER'S YEAST

	Total units	Specific activity units/mg protein	Yield %
First extract*	55,000,000	3,700	
After heat	62,000,000		100
Acetone ppt.	56,000,000	55,000	90
First crystalline preparation	38,000,000	120,000	61
Recrystallized preparation	26,000,000	158,000	42

* Prepared from 200 g of dried yeast. The activity measurements of the first extract are usually low, probably owing to side reactions during the test."

2. *Assay Method*

"*Definition of unit and specific activity.* One unit of enzyme is defined as that amount which causes a change in optical density of 0.001 per minute under the given conditions. Specific activity is expressed as units per milligram of protein. Protein is determined spectrophotometrically and with crude preparations by the biuret reaction.

"*Enzyme.* Dilute stock solution of enzyme in 0.1% bovine serum albumin in 0.01 M potassium phosphate, pH 7.5, to obtain about 200 to 1000 units of enzyme per ml.

"*Assay procedure.* Place 2.2 ml of distilled water, 0.5 ml of pyrophosphate buffer, 0.1 ml of ethanol, and 0.1 ml of DPN into a quartz cell having a 1-cm light path, and start the reaction by addition of 0.1 ml of enzyme. The control cell contains all reagents except the substrate. The first density reading is taken 15 seconds after addition of the substrate, and further readings are recorded at 15-second intervals. The increment in density between the 15- and 45-second readings times 2 is taken as enzyme activity per minute."

TREATMENT OF DATA

In your notebook include a flow sheet showing the principal steps in the procedure, and make a table for your data similar to Table XII.

QUESTIONS

1. Briefly describe three major techniques of enzyme purification that were not employed in this experiment.

2. Name important methods for bringing into solution enzymes from animal, plant, bacterial, and mold cells.

3. Explain the reasons for these steps in the procedure: heating to 55°, addition of acetone in two parts, need for low temperature upon addition of acetone, dialysis, addition of $(NH_4)_2SO_4$ in small portions, need for recrystallization.

4. Discuss briefly the advisability of carrying out two steps of enzyme purification in sequence in exactly the same way except that the second step is in a smaller volume.

5. Suggest another possible assay for alcohol dehydrogenase and compare its merits with the one actually employed.

6. What criteria are used to determine the purity of an enzyme? Are crystalline enzymes necessarily very pure?

Notes on the Choice of an Enzyme for Purification

Although at least one hundred enzymes have been highly purified to date, relatively few seem suitable for a classroom experiment. Most purifications are too lengthy, and a few are so simple that they do not introduce many of the available techniques. The following enzymes have been successfully purified in laboratory courses: see Methods in Enzymology (10) for details.

1. Aldolase and glyceraldehyde-3-phosphate dehydrogenase are satisfactorily prepared but do not provide experience in using varied techniques.

2. Chymotrypsinogen, trypsin, ribonuclease, and deoxyribonuclease are satisfactorily prepared and may all be made from a single source, the pancreas.

3. Catalase preparation from liver is short but is not a typical enzyme preparation.

4. The crystallization of pepsin from commercial pepsin is satisfactory.

5. Isolation of fumarase is somewhat lengthy but works very well.

6. Isolation of enolase is also lengthy, and the alcohol fractionation must be done carefully to avoid loss of activity.

References

1. Schwimmer, S., and Pardee, A. B. 1953. Principles and Procedures in the Isolation of Enzymes. Advances in Enzymology, 14, 375—409. Interscience Publishers. New York.

2. Taylor, J. F. 1953. The Isolation of Proteins. In The Proteins, I, 1—85. H. Neurath and K. Bailey, editors. Academic Press. New York.

3. Peterson, E. A., and Sober, H. A. 1956. Chromatography of Proteins I. Cellulose Ion-Exchange Adsorbents. J. Am. Chem. Soc., 78, 751—5.

4. Reichard, P., and Hanshoff, G. 1956. Aspartate Carbamyl Transferase from Escherichia Coli. Acta Chem. Scand., 10, 548—66.

5. Golding, H. B. 1950. Centrifuging. In Technique of Organic Chemistry, III, 143—70. A. Weissberger, editor. Interscience Publishers, New York.

6. Cummins, A. B. 1950. Filtration. In Technique of Organic Chemistry, III, 485—603.

7. Stauffer, R. E. 1950. Dialysis and Electrodialysis. In Technique of Organic Chemistry, III, 313—61.

8. Tipson, R. S. 1950. Crystallization and Recrystallization. In Technique of Organic Chemistry, III, 363—484.

9. Racker, E. 1955. Alcohol Dehydrogenase from Baker's Yeast. In Methods in Enzymology, I, 500—503. S. P. Colowick and N. O. Kaplan, editors. Academic Press. New York.

10. Methods in Enzymology, Vols. I and II. S. P. Colowick and N. O. Kaplan, editors. 1955—56. Academic Press. New York.

EXPERIMENT 20. ENZYME KINETICS * (2 periods)

OBJECTIVE

This experiment is intended to provide experience in the determination of the equilibrium constant and in measurement of the rate of an enzymic reaction as a function of pH and substrate concentration.

PRINCIPLES

A knowledge of enzyme kinetics is clearly important to anyone who wishes to assay enzyme activity. Beyond this, the study of enzyme kinetics is a useful means of investigation of the mechanism by which enzymes combine with their substrates (1). The quantities most commonly varied in such studies are substrate or cofactor concentrations, pH, temperature, and concentrations of inhibitors. The rate of reaction as a function of substrate and enzyme concentration is represented by the Michaelis-Menten equation (1, 2).

$$v = \frac{k_3(E)(S)}{K_m + (S)}$$

in which v is the rate of reaction at enzyme concentration (E) and substrate concentration (S). From a determination of the rates of reaction at various substrate concentrations one can calculate both the degree of saturation of enzyme by substrate (measured by K_m, the Michaelis-Menten constant) and the maximum rate (k_3E). By studying the effects of variable pH on K_m and k_3, some knowledge of ionizable groups of the enzyme is obtained (1, 3). Interpretation of kinetic data is not easy, and superficial experiments are likely to lead to incorrect conclusions as a result of inadequate consideration of all the variables which affect the system.

The Beckman spectrophotometer will be used to measure the quantity of reduced DPN (DPNH) produced by the reaction of alcohol with oxidized DPN (DPN^+) in the presence of alcohol dehydrogenase. DPNH can be determined readily because it absorbs light at 340 mμ, while DPN^+ does not.

PRINCIPAL EQUIPMENT AND SUPPLIES

Beckman spectrophotometer and 4 cuvettes
Stop watch
Small package of dry bakers' yeast, or purified alcohol dehydrogenase (e.g., from previous experiment)
Ethyl acetate
0.05 M pyrophosphate buffers: 11.2 g $Na_4P_2O_7 \cdot 10H_2O$ plus 375 ml water. Adjust 75 ml aliquots to pH 7.5, 8.0, 8.5, 9.0, and 9.5 with 1 N NaOH, then dilute each to 100 ml

---

* This experiment was derived from one used at the Department of Biochemistry, University of Wisconsin, Madison, Wisconsin.

50 ml of 2×10^{-4} M DPN (mol wt DPN·4H$_2$O = 735). Note that commercial DPN varies in purity as stated on the container's label

0.2 M ethanol (3.0 ml 95% ethanol in 250 ml H$_2$O)

0.2 M semicarbazide (0.22 g semicarbazide·HCl (mol wt = 111.5) plus 7.5 ml water; adjust to pH 8.0 with 1 N NaOH and dilute to 10 ml)

5% acetaldehyde

PROCEDURE

1. *Preparation of Crude Dehydrogenase*

To 7 g of active dry yeast add 15 ml of water and let the mixture stand 30 min at room temperature. Add 2 ml of ethyl acetate, stir for a minute, let stand 20 min, then add 20 ml of water. Centrifuge for 10 min at about 5000 rpm, pour off the supernatant fluid and adjust the pH of the fluid to 8.0. This solution of crude enzyme will keep for a day or two in the cold. If the crystalline enzyme is available, make a solution of 2.0 μg/ml of enzyme in 0.01 M phosphate buffer of pH 7.5 that contains 1 mg/ml of serum albumin to prevent denaturation.

2. *Effect of pH on Rate and Equilibrium Constant of the Alcohol Dehydrogenase Reaction* (4)

Prepare at room temperature the seven reaction mixtures described in Table XIII. By means of preliminary experiments, determine the concentration of enzyme that will bring a mixture like 4 in the table below to equilibrium in 10 min. Try a threefold dilution first.

TABLE XIII. REACTION MIXTURES

	0	1	2	3	4	5	6
0.6 ml P$_2$O$_7$ buffers of the given pH values	7.5	7.5	8.0	8.5	9.0	9.5	8.0
H$_2$O	1.5 ml	0.7	0.7	0.7	0.7	0.7	0.6
2×10^{-4} M DPN	0.8 ml	0.8	0.8	0.8	0.8	0.8	0.8
0.2 M ethanol	0.0 ml	0.8	0.8	0.8	0.8	0.8	0.8
0.2 M semicarbazide	0 ml	0	0	0	0	0	0.1

The equilibrium values for all samples will be determined, and also the approximate rates of DPNH formation for mixtures 3, 5, and 6 will be measured. Add 0.1 ml of enzyme to numbers 0, 1, 2, and 4. Mix each and use number "0" as a blank to adjust the spectrophotometer at 340 mμ. Pour solutions 3, 5, and 6 into spectrophotometer cuvettes, and at 15 sec intervals add 0.1 ml of enzyme, mix each at once and record optical densities of each until the reactions have reached equilibrium. Commence with readings of each sample at 1 min intervals; the optical densities should increase for about 15 min. Then add 0.03 ml of 5% acetaldehyde to the samples in the cuvettes and record the optical densities as a measure of reoxidation of the DPNH. Finally, record the optical densities of the samples 1, 2, and 4 at 5 min intervals until equilibrium is assured, and determine the exact pH of all 7 samples.

3. *Dependence of Rate on Substrate Concentrations and pH* (3)

Devise an experiment similar to the one above, but vary either the concentration of alcohol (0.01, 0.05, and 0.25 M) or DPN$^+$ (1 × 10^{-4}, 5 × 10^{-4}, 25 × 10^{-4} M). Measure the rate of DPNH formation at pH 8.5, and then repeat at pH 8.0 and pH 9.5. Use an amount of enzyme that gives an easily measured initial rate. When the rate of DPN$^+$ reduction has decreased to 10% or less of the initial rate in each case, add 0.03 ml of 5% acetaldehyde and measure the rate of the reverse reaction.

TREATMENT OF DATA

1. Calculate the apparent equilibrium constants for tubes 1 to 5 by the use of the equation

$$K_{apparent} = \frac{(CH_3CHO)(DPNH)}{(CH_3CH_2OH)(DPN^+)}$$

and the relation of the μ moles of DPNH in 3 ml = 0.483 times the optical density. Assume that the acetaldehyde concentration is equal to the DPNH concentration, and that total added DPN is equal to the amount of DPNH found in tube 6 at the completion of reduction. Plot $K_{apparent}$ as ordinate vs. pH on semilog paper. From this curve determine the dependence of the true equilibrium constant on [H$^+$]. Write the equation for the complete reaction catalyzed by yeast alcohol dehydrogenase, and calculate the actual equilibrium constant.

2. At each pH, plot the initial rate (v) against ethanol or DPN concentration (S). Also plot 1/v against 1/(S) (Lineweaver-Burk plot, see ref. [1]), and (S)/v against (S). Such graphs result in straight lines if the Michaelis-Menten equation is obeyed. Calculate K_m and k_3E and plot each quantity against pH. Can you give an interpretation to your results?

QUESTIONS

1. What is the effect of semicarbazide on this reaction? How else could this result be achieved?

2. Calculate the standard free energy change of the alcohol dehydrogenase reaction and the redox potential of DPN from your equilibrium constant and the redox potential of alcohol. What is the effect of pH on this value?

3. A commercial sample of "pure DPN" is assayed by reduction by alcohol dehydrogenase at pH 8.0, reduction by alcohol dehydrogenase at pH 10.0, and reduction by hydrosulfite at pH 8.0. The results show a DPN content of 30, 60, and 90%, respectively. Provide an explanation.

4. If a reaction occurs more slowly at higher than at lower pH, how can you tell if enzyme denaturation is responsible?

5. What are the structural formulas of DPN^+ and DPNH?

References

1. Alberty, R. A.   1956.   Enzyme Kinetics. Advances in Enzymology, 17, 1–64. Interscience Publishers. New York.
2. Neilands, J. B., and Stumpf, P. K.   1955.   Outlines of Enzyme Chemistry, Chap. 7. John Wiley and Sons. New York.
3. Nygaard, A. P., and Theorell, H.   1955.   The Reaction Mechanism of Yeast Alcohol Dehydrogenase (ADH) Studied by Overall Reaction Velocities. Acta Chem. Scand., 9, 1300–1305.
4. Racker, E.   1950.   Crystalline Alcohol Dehydrogenase from Baker's Yeast. J. Biol. Chem., 184, 313–9.

EXPERIMENT 21. ISOLATION OF A COENZYME: ADENOSINE TRIPHOSPHATE (3 periods)

OBJECTIVE

This experiment is intended to give you experience with the isolation of an important coenzyme, adenosine triphosphate, from rabbit muscle.

PRINCIPLES

Coenzymes are naturally occurring compounds that are required for the action of certain enzymes. They serve as agents for the transfer from one enzyme to another of electrons, hydrogen, phosphate, acetyl groups, etc. Coenzymes are of molecular weights 500 to 1000, usually contain ionizable groups, and are found in concentrations of only a fraction of a milligram per gram of biological material. The two principal methods of isolation of coenzymes make use of differential solubility of the compounds or their metal salts and chromatography on ion exchange resins or charcoal. Although coenzymes are not so readily destroyed by chemical means as are enzymes, they are often susceptible to damage from enzyme action; therefore it is usually desirable to separate them quickly from destructive enzymes in crude extracts of tissues.

PRINCIPAL EQUIPMENT AND SUPPLIES

Hypodermic syringe
Sharp knife and scissors
Balance (to weigh several kg)
Blender
Cheesecloth
Rubber gloves
Ice
One rabbit
700 Ml of cold freshly prepared 10% trichloracetic acid (TCA)
700 Ml of cold freshly prepared 5% TCA
50 Ml of 50% (w/v) $MgSO_4 \cdot 7H_2O$
10 Ml of 40% (w/v) NaOH
0.05% bromthymol blue
50 Ml of 2 M Ba(Acetate)$_2$ (mol wt = 255.5)
1 Liter of $\overline{0.2}$ N HNO$_3$
Lohmann's reagent: 100 g Hg(NO$_3$)$_2 \cdot$8H$_2$O in 25 ml concd. HNO$_3$ to which 25 ml water is added
H$_2$S generator or tank
95% ethanol
Ether

PROCEDURE (1, 2)

Inject a rabbit intraperitoneally with 1 ml of 50% MgSO$_4 \cdot$7H$_2$O per kg body weight and then with 0.5 ml every 10 min until the animal no longer gives an eye reflex. Kill the rabbit with a sharp blow on the neck and bleed it at the jugular vein. Skin it by cutting the skin around each hind foot, then join the incisions with a cut up the inside of the legs, and pull the skin forward over the head, trimming as required. Rapidly remove the large muscles (about 500 g) from the hind legs and back, place them in a previously weighed container of cracked ice and reweigh. Quickly grind the muscles in small portions (1 min per portion) with an equal volume of 10% TCA in the blender at 0° to 5°. At once, squeeze the extract through cheesecloth (use rubber gloves) and treat the residue again with an equal volume of 5% TCA in the blender. Filter the combined extracts with suction. Add a few drops of bromthymol blue to the clear filtrate and adjust to pH 6.8 by titrating to a clear green with about 40 ml of 40% NaOH per liter of extract. Precipitate the ATP, along with some other phosphate compounds, with an excess of $Ba(Acetate)_2$ (3.0 ml of 2 $\underline{M}$ $Ba(Acetate)_2$ per 100 g of muscle). Store the mixture in the cold until the next period.

Decant as much supernatant fluid as possible and centrifuge residue to obtain the precipitate. Discard the supernatant fluid and dissolve the precipitate in 0.2 $\underline{N}$ HNO_3 (30 to 50 ml per 100 g of muscle). Filter off a small insoluble residue; then add Lohmann's reagent, using a slight excess (0.6 to 1.0 ml per 100 g of muscle) to precipitate ATP as the Hg salt. After chilling 15 min, centrifuge the mixture, discard the supernatant, and resuspend the precipitate in about 30 ml of water. Saturate the solution with H_2S (in the fume hood!) and shake well to decompose the mercury salt of ATP. Centrifuge the mixture and save the supernatant fluid. Wash the HgS with 0.2 $\underline{N}$ HNO_3 and discard the precipitate. Aerate the combined supernatant and washings in the hood until excess H_2S has been removed. Neutralize the solution to pH 6.8 and add $Ba(Acetate)_2$ in slight excess, until an additional drop will not cause more precipitate to form. The preparation may be chilled and stored until the next period.

The precipitate is collected by centrifugation and redissolved in a small volume of 0.2 $\underline{N}$ HNO_3. A little H_2SO_4 is added to carry down traces of HgS with a small precipitate of $BaSO_4$ upon centrifugation. The ATP may then be further purified and stored in one of two forms.

a. *As the barium salt.* The solution is neutralized, excess Ba(Acetate)$_2$ is added, and the solution is chilled 30 min and then centrifuged in glass containers. The wet precipitate is washed by centrifugation with 6 to 8 times its volume of 1% $Ba(Acetate)_2$, then with 50% ethanol, 75% ethanol, 95% ethanol, and, finally, with ether. The precipitate is then spread over the bottom of the centrifuge tubes and dried under vacuum. The yield should be about 1.5 g of $Ba_2ATP \cdot 4H_2O$ per 500 g of muscle.

b. *As the crystalline sodium salt.* Mix 1 g of barium ATP with 4 ml of 1.5 $\underline{N}$ HNO_3. Add 0.7 ml (a slight excess) of saturated Na_2SO_4 and centrifuge off the precipitate of $BaSO_4$. Wash the precipitate with 1 ml of 0.1 $\underline{N}$ HCl and add the washing to the supernatant solution. Adjust the pH of the solution to about pH 3.5 with 40% NaOH. Add 0.8 ml of 95% ethanol per ml of solution, seed with crystals of ATP if available, and leave the solution at room temperature for 1 hr to crystallize. Wash the centrifuged material twice with 67% alcohol and twice with 95% alcohol and dry the crystals in vacuo at room temperature.

TREATMENT OF DATA

In your notebook include a flow sheet of the process. Indicate points at which various other muscle extractives and salts are removed.

QUESTIONS

1. What are the roles of ATP in metabolism? Why would you expect muscle to be a good source of ATP?

2. Why is magnesium sulfate anesthesia used, and why must the muscle be cooled and ground rapidly?

3. Suggest another method for isolation of ATP.

4. Barium ATP can provide a cofactor for a number of reactions for which crystalline ATP is inactive. Suggest an explanation. How would you test your hypothesis?

ALTERNATIVE EXPERIMENTS, 21

Several coenzymes and substrates are suitable for purification by the class in the laboratory. The procedures for most of these preparations may be found in refs. (1) or (2), or in one of the volumes of <u>Biochemical Preparations</u>.

1. DPN can be purified fairly readily, although the procedure is more laborious than that for ATP and the quantity of product obtained is smaller. The final chromatographic purification is lengthy.

2. Cytochrome c can be readily purified, although the final chromatographic purification is time-consuming.

3. Glucose-1-phosphate can be prepared fairly readily.

4. 3-Phosphoglyceric acid is not difficult to isolate.

5. Preparations of TPN, coenzyme A, or flavin coenzymes are considerably more difficult.

References

1. Umbreit, W. W., Burris, R. H., and Stauffer, J. F. 1957. <u>Manometric Techniques,</u> 3rd ed., Chap. 17. Burgess Publishing Co. Minneapolis, Minn.
2. Berger, L. 1957. Isolation of ATP from Muscle. In <u>Methods in Enzymology,</u> III, 862–864. S. P. Colowick and N. O. Kaplan, editors. Academic Press. New York.
3. Berger, L. 1956. Crystallization of the Sodium Salt of Adenosine Triphosphate. Biochim. et Biophys. Acta, <u>20</u>, 23–6.

EXPERIMENT 22.   PROPERTIES OF ADENOSINE
TRIPHOSPHATE (2 periods)

OBJECTIVES

The experiment is designed to illustrate some techniques used in the study of coenzymes and to investigate the chemical properties and purity of a sample of ATP.

PRINCIPLES

Tests for composition and purity of coenzymes are of value to ascertain that one is working with pure compounds, and also to aid in determination of unknown structures. ATP readily lends itself to study by a variety of chemical, physical, and biochemical techniques of wide applicability (1), because of its composition, charge, absorption spectrum, and role as a coenzyme. The techniques applied are useful in the study of other coenzymes and metabolites.

EQUIPMENT AND SUPPLIES

Beckman spectrophotometer and 4 silica cuvettes
Water bath
Jar for ascending chromatography
Ultraviolet lamp
10 Ml of a 1% solution of the sodium salt of ATP; preferably use the material prepared in Experiment 21
Fiske-Subbarow reagents (see Appendix III)
Orcinol (recrystallized from benzene)
0.1% $FeCl_3$ in concentrated HCl
Dowex 1 (Cl^-) (200–400 mesh; 8X)
Isobutyric acid
Hexokinase
Glucose-6-phosphate dehydrogenase (Zwischenferment)
TPN

PROCEDURES

Perform as many of the experiments as time permits; choose those which provide you with new techniques.

1. *Chemical Tests*

a. *Absorption of ultraviolet light.* Ultraviolet light absorption by ATP is primarily due to the adenine group. The molar extinction coefficient of ATP at pH 7.0 at 259 mμ is 15.4 $\times$ 10^3. Determine the absorption spectrum from 220 to 310 mμ of your ATP preparation, and calculate its ATP content from the optical density (OD) at 259 mμ. A comparison of OD ratios at 250/260 mμ and 280/260 mμ is also instructive (see Table VI, Experiment 5).

b. *Total phosphate and acid-labile phosphate.* Determine phosphate by the Fiske-Subbarow method (Appendix III) on duplicate aliquots of your solution of ATP before hydrolysis, after 10 min in 1 N HCl at 100°, and after complete hydrolysis. Compare color values of the samples with those of a standard curve prepared from aliquots of a standard phosphate solution. From these data calculate the purity of your ATP.

c. *Ribose content* (2). Mix, in duplicate, 2 ml of ATP solution (containing about 0.07 mg of ATP) with 2 ml of 1% orcinol freshly dissolved in 0.1% $FeCl_3$ in concentrated HCl. Cover each test tube with a marble and heat for 30 min in a boiling water bath, then cool, dilute to 4.0 ml, and read optical densities of the samples at 660 mμ. Compare values with those of standard ribose, xylose, or arabinose samples assayed at the same time.

2. *Chromatographic Tests for Purity of ATP:*

a. *Ion exchange chromatography* (3). (Consult Experiment 5 for more complete details of the use of ion exchange resins.) Pour a bed 1 cm high in a chromatography column of 1 cm diameter of Dowex 1 (Cl$^-$) (200 – 400 mesh; 8X). Wash the resin with 1 M HCl until the optical density at 259 mμ of the wash solution is less than 0.1, and then with water until the wash is nearly neutral. Apply about 5 mg of your ATP in 1 ml of 1 M NH_4OH to the column. Elute in succession with 20 ml of water, 60 ml of 0.003 N HCl, 60 ml of 0.01 N HCl plus 0.02 M NaCl, 100 ml of 0.01 N HCl plus 0.2 M NaCl, and finally with 20 ml of 1 N HCl. Do not allow the column to become dry during these operations. The flow rate should be about 3 ml per min. Collect the effluent in 20 ml portions and read optical densities at 259 mμ. Plot OD vs. ml of eluate, and calculate the composition of your ATP according to the method shown in ref. (3).

b. *Paper chromatography* (4). Apply 0.05 to 0.1 mg of your 1% ATP, in solution, as a 2 to 3 mm diameter spot placed $1\frac{1}{4}$ in. from the end of a strip of Whatman No. 1 paper at least 12 in. long. Also apply spots of ATP, ADP, and AMP standards. Run an ascending chromatogram for 16 hr with a solvent system consisting of 1 ml of conc'd. ammonia in 33 ml of water and 66 ml of isobutyric acid. Mark the solvent front, dry the paper, and outline the spots found under UV light in a dark room. The R_f (distance moved by spot divided by distance moved by solvent) of ATP = 0.20, ADP = 0.30, and AMP = 0.45.

Alternatively, paper electrophoresis of ATP may be tried (5) instead of paper chromatography.

Biochemical Assay for ATP (6)

The assay depends on the reactions:

$$\text{ATP} + \text{glucose} \xrightarrow{\text{hexokinase}} \text{ADP} + \text{glucose-6-}PO_4 + H^+$$

$$\text{glucose-6-}PO_4 + TPN^+ + H_2O \xrightarrow{\text{Zwischenferment}} \text{6-phospho-gluconolactone} + \text{TPNH} + H^+$$

Reduced TPN is measured by its absorption at 340 mμ. This technique of measuring a product of an enzyme reaction that is difficult to analyze by coupling it with a second reaction whose product is readily analyzed is commonly used.

Assay procedure. Mix 0.2 ml of 0.5 M glucose, 0.2 ml of 0.15 M MgCl$_2$, about 100 μg of hexokinase (add more if the reaction is slow), 0.1 ml of Zwischenferment (3 mg/ml), 0.1 ml of TPN (2 mg/ml), 2.0 ml of 0.05 M phosphate (pH 7.5), and water to make a final volume, after ATP addition, of 3.0 ml. Read the optical density at 340 mμ. The sample of ATP solution, containing about 0.1 mg of ATP, is added to start the reaction. Read the optical density at 340 mμ at the completion of reaction, as determined by no further change of optical density with time. The molar extinction coefficient of reduced TPN is 6.22×10^3 at 340 mμ. Less than 1 mole of TPNH is produced per mole of ATP added; presence of 0.002 M CN$^-$ increases the yield of TPNH (D. R. Sanadi, personal communication).

Alternatively, the firefly luminescence assay may be tried if the necessary materials are available (1).

TREATMENT OF DATA

In your notebook include calculations from the various experiments, and state what they show regarding chemical composition and biological and chromatographic purity of your ATP preparation.

QUESTIONS

1. Which phosphate groups are removed from ATP by heating 10 min at 100° in 1 N HCl?

2. How was the structure of ATP originally established?

3. Write the structural formulas of the compounds involved in the hexokinase and Zwischenferment reactions.

4. Describe another method for estimation of the hexokinase reaction. What are the relative merits of your suggested method and the assay used in this experiment?

5. A bottle is labeled 100 mg of ADP. How would you determine the purity of its contents?

6. Give examples of other coupled reactions used for enzyme assays or substrate determinations.

References

1. Strehler, B. L., and Totter, J. R. 1954. Determination of ATP and Related Compounds: Firefly Luminescence and Other Methods. In Methods of Biochemical Analysis, I, 341−56. D. Glick, editor. Interscience Publishers. New York.
2. Umbreit, W. W., Burris, R. H., and Stauffer, J. F. 1957. Manometric Techniques, 3rd ed., Chap. 16. Burgess Publishing Co. Minneapolis, Minn.
3. Cohn, W. E., and Carter, C. E. 1950. The Separation of Adenosine Polyphosphates by Ion Exchange and Paper Chromatography. J. Am. Chem. Soc., 72, 4273−5.

4. Berger, L. 1956. Crystallization of The Sodium Salt of Adenosine Triphosphate. Biochim. et Biophys. Acta, 20, 23–26.
5. Hilz, H., and Lipmann, F. 1955. The Enzymatic Activation of Sulfate. Proc. Nat. Acad. Sci. US, 41, 880–90.
6. Kornberg, A. 1950. Reversible Enzymatic Synthesis of Diphosphopyridine Nucleotide and Inorganic Pyrophosphate. J. Biol. Chem., 182, 779–93.

EXPERIMENT 23. ENZYMATIC RESOLUTION OF AMINO ACIDS (2 periods)

OBJECTIVES

This experiment is intended to illustrate the utility of certain enzymes, even in the impure state, as preparative reagents, to illustrate the specificity of enzymes, and to provide experience in the operation of the polarimeter.

PRINCIPLES

Enzymes can be used as specific catalysts for various preparative and analytical processes, both in research and in industry. As an example, the utility and also the specificity of the enzyme acylase is illustrated in the present experiment. The acylase reaction provides an excellent means of preparing pure optical isomers of amino acids. When an N-acetyl-DL-amino acid is exposed to acylase, only the L-isomer is hydrolyzed. Subsequent separation of the free L-amino acid from the acetylated D-form is easily accomplished by virtue of the lower solubility of the former in alcohol.

Measurements with the polarimeter of changes in optical rotation have been used in the past for assaying enzyme activities. The classic example is the assay of invertase (1). Polarimetric methods generally require large amounts of enzymes and substrates and are limited to reactions in which a considerable change of rotation per mole occurs. The development of photoelectric polarimeters (2) probably will greatly increase the usefulness of this technique in Enzymology.

PRINCIPAL EQUIPMENT AND SUPPLIES

Blender
Polarimeter
Refrigerated centrifuge
Vacuum distillation apparatus
3 Fresh-frozen hog kidneys (or commercial acylase powder)
N-acetyl-DL-methionine (4 g)
2 N HCl
2 N NaOH
Decolorizing charcoal
95% ethanol

PROCEDURE

1. *Preparation of Acylase* (3)

Three fresh-frozen hog kidneys (about 250 g) are thawed, defatted, and homogenized for 2 min in a blender with 2 vols. of ice water. The homogenate is centrifuged at about 10,000 × g for 15 min, and the supernatant solution is chilled to $0°$ on ice. The solution is brought to pH 4.7 by careful addition of 2 N HCl, and the resulting thick suspension is immediately centrifuged at $0°$ at 10,000 × g for 10 min. The clear, red supernatant fluid is quickly adjusted to pH 6.5 by addition of 2 N NaOH, and 266 g of solid $(NH_4)_2SO_4$ is added per liter of solution, whereupon the pH decreases to 6.0. The material is centrifuged 10 min at 10,000 × g and the supernatant solution is discarded. The sediment is suspended in 4 ml of ice water, and dialyzed against running tap water until completely free of $(NH_4)_2SO_4$ when tested with Ba^{++}. The contents of the dialysis sack is centrifuged to remove inactive protein, and 0.5 ml of the supernatant solution is used as the enzyme in the experiment below; the remainder of the preparation may be stored frozen or lyophillized. The enzyme may be purified further if desired (3).

2. *Enzymatic Hydrolysis of N-Acetyl-L-Methionine* (4)

Dissolve 4.0 g of N-acetyl-DL-methionine in 150 ml of water and add concd. ammonia to bring it to pH 7.5 to 8.0. Add 0.5 ml of your acylase solution and incubate the mixture overnight at room temperature.

3. *Purification of L-Methionine*

Bring the reaction mixture to pH 4.5 with about 3 ml of acetic acid, add 100 mg of decolorizing carbon, heat over an open flame for a few moments to coagulate the protein, and filter with suction. Place the filtrate in a 500 ml round-bottom flask and evaporate off most of the solvent under vacuum at $35°$ or less. Add 25 ml of benzene and evaporate the solution to dryness. Dissolve the residue in 5 to 10 ml of water, add 30 ml of ethanol, chill for 30 min or more, and filter off the precipitate. Dissolve the precipitate in a small volume of hot water, add hot ethanol to produce a faint turbidity, and cool gradually to crystallize the L-methionine. The yield should be 50 to 75%. Establish purity of your product by its specific optical rotation ($[\alpha]_D^{20} = +23.4°$ in 3 N HCl). It should be possible to recover N-acetyl-D-methionine from the alcoholic solutions if desired (4).

TREATMENT OF DATA

Calculate the yield and purity of your product.

QUESTIONS

1. What enzymatic and non-enzymatic procedures for resolution of optical isomers are known?
2. Why is the removal by acid precipitation of impurities from acylase performed rapidly?
3. List some highly specific enzymes and other quite non-specific enzymes.
4. What is the principle of the Van-Slyke ninhydrin-CO_2 method of analysis (3)? Compare it with the colorimetric ninhydrin method (see Experiment 10).

References

1. Hestrin, S., Feingold, D. S., and Schramm, M.   1955.   Hexoside Hydrolases. In Methods in Enzymology, I, 231−57. S. P. Colowick and N. O. Kaplan, editors. Academic Press. New York.
2. Levy, G. B., and Cook, E. S.   1954.   A Rotographic Study of Mutarotase. Biochem. J., 57, 50−5.
3. Birnbaum, S. M.   1955.   Aminoacylase. In Methods in Enzymology, II, 115−19. S. P. Colowick and N. O. Kaplan, editors. Academic Press. New York.
4. Greenstein, J, P.   1957.   Resolution of DL-Mixtures of α-Amino Acids. In Methods in Enzymology, III, 554−570. S. P. Colowick and N. O. Kaplan, editors. Academic Press. New York.

EXPERIMENT 24. OPERATION AND CALIBRATION OF THE WARBURG RESPIROMETER (1 period)

OBJECTIVE

This experiment is intended to familiarize you with the Warburg respirometer.

PRINCIPLES

The Warburg respirometer is a sensitive device for measurement of gas pressure changes; it can be used to follow reactions in which a gas is evolved or taken up. Such reactions are numerous in biochemistry; examples are reactions catalyzed by oxidases, decarboxylases, or, in the presence of bicarbonate, by enzymes which produce acid. Measurements can be made with a sensitivity of one micromole and an accuracy within a few per cent. A prime advantage of the respirometric method is that it may be used to make determinations as a function of time. This can be done in a complex assay mixture such as a tissue extract: extraneous materials in solution do not interfere with the final measurement as they do in colorimetric assays. A second advantage of respirometry is that a great deal of data may be obtained in a relatively short time, since as many as nineteen separate samples can be measured simultaneously.

With the Warburg apparatus, gas evolution or uptake leads to a pressure change in the apparatus at constant volume, and the pressure change is measured with a manometer filled with an aqueous solution (Brodie's solution, see ref. [1]). The pressure change (Δh) read in the manometer can be converted into the amount of gas exchanged (Δx) by the equation $\Delta x = k\Delta h$. The "flask constant" k depends primarily on the gas volume of the flask and the manometer in which the reaction occurs.

PRINCIPAL EQUIPMENT AND SUPPLIES

Warburg respirometer (bath at $37°$)
10 Warburg flasks with side-arms
10 Ml of 0.5 N NaOH
0.0100 M pyrogallol (make a fresh solution of 126 mg of pyrogallol in 100 ml of water plus 4 drops of glacial acetic acid)

PROCEDURE

A condensed list of operating instructions is presented below, but it is necessary to read one of the references (1, 2) to learn the details of operation.

1. *Check-List for Warburg Respirometer Operation*

a. Draw up a table for recording data in your notebook. The form on the next page has proven satisfactory.

Time, t	Δt min	Thermoba-rometer	ΔTB	Flask 1 (k = 1.10)			Flask 2, etc.
				h	Δh*	O_2/10 min	
2:42	0	Start					
2:52	10	145		271			
3:02	20	145	0	253	-18	20	
3:12	30	150	$+5$	238	-20	22	
3:22	40	149	-1	219	-18	20	

* Pressure change in mm corrected for thermobarometer change (ΔTB).

    b. Turn the water-bath heater on well in advance.

    c. Put 3 vertical strips of light grease on each manometer joint (stiff grease will permit creeping of the flasks on the joint).

    d. Grease rims of flask center-wells if KOH is to be used.

    e. Put stable substances into main compartments and side-arms of flasks, and KOH in center-wells.

    f. Put flasks on ice if the enzyme is unstable.

    g. Put labile substances into body or side-arm of flasks.

    h. Insert filter paper rectangles (preferably Whatman No. 42), accordion folded, in center-wells to increase surface area of the KOH solution.

    i. Attach side-arm stoppers, using light grease.

    j. Fasten flasks to manometers with rubber bands or springs.

    k. Change the gas in the flasks if required. Flasks may be gassed after they are placed in the bath if the enzyme is stable.

    l. Turn on the shaker and transfer each flask and manometer to the bath at 20 sec intervals.

    m. After 5 min carefully tighten the connections of flasks to manometers by rotating the flasks with gentle pressure. Oppose the pressure of one hand with that of the other, rather than push against the glass. Set the manometers so that when the stopcocks are closed and the reading of the closed arm is adjusted to the standard position (150 mm), fluid in the open arm will be near 300 mm (for gas uptake) or near 0 mm (for gas evolution).

    n. Ten or 15 min after the flasks are immersed in the bath, set the closed arm of the first manometer at 150 mm and record the reading of the open arm. Continue by reading the other manometers in the same way at 20 sec intervals.

    o. When it is desired to mix the side-arm contents with the flask contents, usually after one or two readings, remove flask and manometer from the bath and tilt the assembly back and forth carefully until the solutions run together. (Close off the open end of the manometer with your finger to prevent the Brodie's solution from being drawn into the flask.)

    p. Continue reading each manometer, usually at 10 min intervals, until the desired data are obtained.

    q. Finally, open the stopcocks, remove manometers, take off flasks, wipe the manometer joints, turn off heater and shaker, and check pH of the flasks (with a drop of phenol red or other indicator).

    r. Clean flasks are essential. Rinse flasks with very hot water to remove grease and place them in a solution of 1 tablespoon of Na_3PO_4 per

2 liters of water and boil the solution gently for 15 min. Rinse each flask four times with tap water and then four times with distilled water. More rigorous cleaning is achieved by heating the flasks in a sulfuric-nitric acid bath.

2. *Calibration of Warburg Flasks*

The flasks will be calibrated by a method which illustrates the usual mode of operation of the respirometer.* Pipet 1.00 ml of pyrogallol solution into the main part of each of 10 flasks and 0.30 ml of 0.5 N NaOH into each side-arm. Use an extra flask (containing 1 ml of water) for a thermobarometer. Put the flasks into the apparatus (at 37°) as described above and set the open sides of the manometers near 300. Make several readings to be sure equilibrium is achieved. Then tilt the alkali into the pyrogallol. Make several readings of the new pressures, and the bath temperature.

TREATMENT OF DATA

Calculate the flask constants for O_2 and CO_2 at fluid volumes of 3.2 and 3.0 ml, respectively, at 37° for each flask, as follows:

1. Subtract the average final manometer reading from the initial one for each flask. Correct this pressure change for the change of the thermobarometer.

2. Assuming that one mole of pyrogallol takes up 33.6 liters of O_2 (1.5 moles) at standard temperature and pressure, calculate the flask constant under the experimental conditions.

3. Calculate the total volume in ml (V) of flask plus manometer to the 150 mm mark from the equation

$$V = 1.27 + 3820/h \qquad [1]$$

where h is the pressure change in mm Brodie's solution.

QUESTIONS

1. What other methods are available for calibration of Warburg flasks? Compare their merits with those of the present method.

2. Derive equation [1] used in this experiment.

3. What is the structure of pyrogallol?

4. Compare the sensitivities of typical assays involving titration, spectrophotometry ($DPNH^+$), and respirometry.

---

* This method of calibration is credited to H. T. Gordon and M. S. Mulla, 1955, unpublished.

References

1. Umbreit, W. W., Burris, R. H., and Stauffer, J. F. 1957. Manometric Techniques, 3rd ed., Chaps. 1 to 5. Burgess Publishing Co. Minneapolis, Minn.
2. Dixon, M. 1951. Manometric Methods, 3rd ed., 9–24. Cambridge University Press. Cambridge, England.

EXPERIMENT 25.   ENZYME INDUCTION (1 period)

OBJECTIVES

This experiment provides experience in culture of bacteria and in measurement of oxygen uptake with the Warburg respirometer. Respiration will be measured under conditions where the bacteria form inducible enzymes.

PRINCIPLES

Bacteria possess a number of advantages over plant and animal tissues for metabolic studies. A bacterial culture provides a uniform, reproducible, easily prepared mass of cells which may be handled readily and subdivided into equal portions for comparison under different conditions. The metabolic activity per unit mass of bacteria is high and the choice of reactions which may be studied is very wide. However, masses of bacteria greater than a few grams are tedious to cultivate and harvest, and this is a disadvantage for work on enzyme isolation from bacteria.

One of the most remarkable properties of bacteria is their ability to synthesize certain enzymes, called adaptive or inducible enzymes, when appropriate substrates or related substances are present in the bacterial growth medium (1). These substances are called inducers. The present experiment provides an example of "sequential induction" (2): an added compound A is metabolized to B by an inducible enzyme (E_A); B then serves as inducer for a second enzyme E_B, which now converts B to C and so on. Demonstration of sequential induction provides evidence for metabolic pathways involving conversion of A to B, to C, etc. It is desirable to obtain confirmatory evidence by other methods; preferably, cell-free extracts are used to demonstrate the individual steps (3, 4).

In this experiment oxygen uptake is used as a measure of the formation of oxidative enzymes in intact bacteria. Any CO_2 evolved is absorbed by a solution of KOH in the center-well. Such measurements are among the simplest which may be made with the Warburg respirometer.

PRINCIPAL EQUIPMENT AND SUPPLIES

Warburg respirometer (bath at $30°$)
10 Warburg flasks with side-arms
Watch (preferably a stop watch)
Autoclave
Pseudomonas fluorescens, strain A.3.12 (American Type Culture Collection number 12633)*
Glutamic acid
30% KOH
Yeast extract

---

*American Type Culture Collection, 2112 M Street, N.W., Washington 7, D. C.

Whatman No. 42 filter paper rectangles 1 in. wide and slightly longer than the height of the center-well. Fold these accordion fashion.

1 liter of 0.02 $\underline{M}$ phosphate buffer, pH 7.0

15 Mg of $\underline{p}$-hydroxy-benzoic acid in 100 ml of the above buffer

15 Mg of benzoic acid in 100 ml of buffer

15 Mg of protocatechuic acid in 100 ml of buffer

15 Mg of catechol in 100 ml of buffer (make this solution a few minutes before use)

0.05% phenol red or bromphenol blue indicator

PROCEDURE

1. *Growth Medium*

Half of the class should make one of these media, and half should make the other. Dissolve 0.04 g of $\underline{p}$-hydroxy-benzoic acid, 0.4 g of glutamic acid, 0.1 g of NH_4NO_3, 0.13 g of $K_2HPO_4 \cdot 3H_2O$, 0.05 g of $MgSO_4 \cdot 7H_2O$, and 0.02 g of yeast extract in 100 ml of distilled water; adjust to pH 7.0 with KOH. Place the medium in a 500 ml flask provided with a gauze-wrapped cotton plug if the flask is to be aerated by shaking. If the flask is to be aerated by bubbling, insert a cotton-plugged glass tube through the flask plug. Sterilize the filled flask by autoclaving for 15 min at 15 lbs steam pressure, and cool it to 30°.

The second medium is the same as the first except that benzoic acid is substituted for $\underline{p}$-hydroxy-benzoic acid.

2. *Growth of Bacteria*

The evening before the experiment, inoculate the flask of medium with 0.3 ml of a highly turbid culture of Pseudomonas fluorescens (prepared by inoculation, from a slant, of a sterile medium lacking aromatic acids); use sterile technique. Aerate the flask at 30° until there are 1 to 2 $\times$ 10^9 bacteria per ml. This opaque, slightly fluorescent suspension will give a reading of 100 to 200 on the Klett colorimeter with green filter (No. 54). Centrifuge the culture 5 min at 5000 rpm and wash the bacteria by resuspending them in 30 ml of 0.02 $\underline{M}$ phosphate buffer. Suspension is easily done by sucking the mixture in and out of a 10 ml pipet several times. Centrifuge the suspension again and resuspend the bacteria in a few ml of buffer, at a concentration such that the turbidity is 1200. (Make the measurement on a suitable dilution.)

3. *Measurement of Oxygen Uptake*

Follow the outline for operation of the respirometer given in Experiment 24. Into four pairs of flasks put 2.7 ml of the $\underline{p}$-hydroxy-benzoate, benzoate, protocatechuate, and catechol solutions and put phosphate buffer into the ninth flask. Center-wells of all flasks are to contain 0.2 ml of KOH and a filter-paper wick. Put 0.30 ml of the concentrated bacterial suspension into the side-arm of each flask. Equilibrate the flasks in the bath at 30° for 10 min, mix bacteria and substrate carefully, and read pressure changes every 5 min for 30 min and then at longer time intervals until enzyme activity is observed in all flasks. Note that it is possible to read beyond the end of the manometer scale (see ref. [5], Chap. 5), although it is often better simply to reset the manometer and recommence measurements with the next reading. Check the pH of each flask with phenol red or bromphenol blue at the end of the experiment: the solutions should be neutral.

TREATMENT OF DATA

Calculate the oxygen uptake per min for each flask in each time interval and plot the rate of uptake against time for each flask. Compare your results with those obtained by members of the class who grew bacteria in the other medium. Write up your results in the form of a note for the Journal of Bacteriology.

QUESTIONS

1. Enzyme determinations are less liable to be influenced by unknown reactions and factors such as permeability if cell-free extracts are used rather than intact cells (3).

(a) What methods for disrupting bacteria are available?

(b) What other assays could you use, particularly with disrupted bacteria, for the enzyme activities studied in this experiment?

2. If a compound is used at once by bacteria previously grown on benzoate, is it necessarily a product of benzoate metabolism? How would you decide?

3. What explanations are possible if a compound is not metabolized by a microorganism?

4. Can you cite an example of the importance of enzyme induction for an enzyme isolation?

ALTERNATIVE EXPERIMENTS, 25

1. A convenient reaction sequence for the study of oxygen uptake by animal tissue is the succinoxidase system (see ref. [5], Chap. 12).

2. Enzyme induction to utilize carbon sources such as glycerol, lactose, galactose, or maltose by Escherichia coli may be studied in the same manner as that used in the above experiment. Use 5×10^9 bacteria per flask in a medium containing 0.1% substrate and inorganic salts (see Experiment 28 and ref. [6]).

References

1. Society for General Microbiology. 1953. Adaptation in Micro-Organisms, 98–
   183 especially. R. Davies and E. F. Gale, editors. Cambridge University Press.
   Cambridge, England.
2. Stanier, R. Y. 1950. Problems of Bacterial Oxidative Metabolism. Bacteriol. Rev.,
   14, 179–91.
3. Stanier, R. Y. 1955. Cleavage of Aromatic Rings with Eventual Formation of β-
   Ketoadipic Acid. In Methods in Enzymology, II, 273–87. S. P. Colowick and N. O.
   Kaplan, editors. Academic Press. New York.
4. Kogut, M., and Podoski, E. P. 1953. Oxidative Pathways in a Fluorescent Pseudomonas. Biochem. J., 55, 800–811.
5. Umbreit, W. W., Burris, R. H., and Stauffer, J. F. 1957. Manometric Techniques,
   3rd ed. Burgess Publishing Co. Minneapolis, Minn.
6. Monod, J., and Cohn, M. 1952. La Biosynthese induite des enzymes (adaptation
   enzymatique). Advances in Enzymology, 13, 67–119. Interscience Publishers. New
   York.

EXPERIMENT 26. RESPIRATORY QUOTIENT OF TISSUE SLICES (1 period)

OBJECTIVES

This experiment provides experience in preparation of tissue slices and in measurement of simultaneous CO_2 production and O_2 uptake.

PRINCIPLES

Tissue slices are used for the study of metabolism in intact animal or plant cells. Both the strength and weakness of the use of tissue slices lie in one property—the ability to catalyze many reactions. Most of the reactions of the intact cell no longer occur when the cell wall is broken as a result of dilution or destruction of cofactors, enzyme denaturation, and probably other causes such as separation of macromolecules. Therefore, the slice is useful to study these reactions. However, a disadvantage of intact cells for metabolic studies is that the details of a single reaction, such as requirements for cofactors, pH optimum, etc., are not readily measured, because the internal conditions of the cells cannot be controlled. A second disadvantage of slices results from the presence of reserve nutrients in the cells: considerable metabolism occurs independently of the presence of substrates (endogenous metabolism) and obscures the reaction under study. The extent of a specific reaction in a tissue slice is usually measured best by some method more specific than oxygen uptake. For example, a colorimetric assay may be used to determine a specific product of the reaction.

Tissues must be cut into very thin slices in order that diffusion of both substrates and O_2 is not the limiting factor in metabolism (1). The technical difficulty of making adequately thin slices limits the use of the method somewhat, and there is evidence that only a limited fraction of the cells in a slice remain in a physiologically favorable condition (2). Other devices for study of metabolism in slices have been described (3).

A number of methods are available for simultaneous measurement of both oxygen uptake and CO_2 production. The "direct method" (1), used in this experiment, requires two flasks. In one flask, CO_2 is absorbed by KOH in the center-well and only the pressure change caused by oxygen uptake is observed. In the second flask KOH is omitted, and the pressure change is the net result of both oxygen uptake and CO_2 release. Data obtained with the first flask are used to calculate the pressure change caused by O_2 uptake in the second flask, and the pressure change due to CO_2 is determined by difference. Note that the flask constants for O_2 and CO_2 of a flask are different, and that the CO_2 constant depends on pH. Also, the flask volumes and amounts of tissue in the two flasks are different and corrections must be made for these differences.

PRINCIPAL EQUIPMENT AND SUPPLIES

One rat (200 – 300 g)
Scissors
Tissue slicer (see ref. [1], Chap. 9)
Watch or stop watch
Warburg respirometer (set at 37°)
10 Warburg flasks
100 Ml of Krebs-Ringer phosphate solution (see ref. [1], Chap. 9)
10 Ml of 10% glucose
30% KOH and center-well papers
10% trichloracetic acid (TCA)
Folin-Ciocalteu reagents (see Appendix II)

PROCEDURE

Kill the rat by placing it in a jar with an ether-soaked rag or by striking its head firmly against the table-top; then cut its backbone at the neck
with scissors and allow the blood to drain out. Open the abdomen, remove
the kidneys, and cool them on ice at once. If you are not familiar with the
locations of the major organs, have someone point them out.

Prepare from the kidneys cubes measuring 0.5 cm on a side and from
these cut 30 or more slices (1). Keep the slices in cold Krebs-Ringer
phosphate solution until they are needed. Add 2.9 ml of Krebs-Ringer
phosphate solution and 3 kidney slices to each of ten Warburg flasks (consult Experiment 24 for details of operation). Put 0.2 ml of KOH into the
center-wells of flasks 1 to 5. Pipet 0.1 ml of 10% glucose into flasks 1 to
3 and 6 to 8, and 0.1 ml of water into the others. Attach the flasks to their
manometers and place them in the bath. Set the manometer fluid in the
open arm near 300 for samples 1 to 5 and near 150 for samples 6 to 10.
Shake the flasks and make at least four readings at 10 min intervals.

At the end of the experiment determine the pH with phenol red and
then the total protein in each flask. For the second measurement, transfer the contents of each flask to a separate centrifuge tube and add to each
an equal volume of 10% TCA and chill for 10 min. Centrifuge, and dissolve each precipitate in 20 ml of 0.5 $\underline{M}$ KOH (3% KOH) with warming. Run
the Folin-Ciocalteu test (see Appendix II) in duplicate, using 0.05 ml of
the KOH extract of each sample. One may, alternatively, blot the slices
and weigh them to obtain an estimate of the amount of wet tissue in each
flask.

TREATMENT OF DATA

From your data calculate the rate of O_2 uptake per mg of protein or
wet tissue in flasks 1 to 5. Calculate the pressure change that this oxygen
uptake should cause in each of the corresponding flasks of set 6 to 10, and
subtract this pressure change from the observed one. Multiply the pressure difference by the flask constant for CO_2 at the pH of the reaction, to
obtain μl of CO_2 evolved. Calculate the respiratory quotient of the slices
with and without glucose present. Record your calculations in your notebook.

QUESTIONS

1. What effect did glucose have on respiration? Why?

2. What effects do buffer concentration, pH, or HCO_3^- concentration have in an experiment of this sort on the pressure change caused by CO_2 release? How could you measure CO_2 release at pH 8.0?

3. Yeast in 0.10 M acetate buffer of pH 5.5 plus 1% glucose was placed in the Warburg respirometer (atmosphere of air and no alkali in the center-well). No pressure change was observed. What conclusions are possible regarding the R.Q. (respiratory quotient) of glucose oxidation? What controls are needed?

ALTERNATIVE EXPERIMENTS, 26

1. The ability of kidney slices to oxidize a radioactive compound such acetate-1-C^{14} may be studied (4). Extra equipment, aside from the labeled compound, includes a Geiger-Mueller counter, planchets, and plating equipment (see Section III, this manual).

2. Instead of performing an experiment with slices, you may wish to study metabolism by mitochondrial particles (5). The oxidation of Krebs cycle intermediates (1) or of octanoate (6) may be measured with the Warburg respirometer.

3. Oxidative phosphorylation may be determined; the system has been well described (7).

References

1. Umbreit, W. W., Burris, R. H., and Stauffer, J. F. 1957. Manometric Techniques, 3rd ed. Burgess Publishing Co. Minneapolis, Minn.
2. Moyson, F. 1956. Limitation de l'emploi des coupes tissulaires pour l'etude des phenomenes metaboliques. Experientia, 12, 103–4.
3. Drabkin, D. L., and Marsh, J. B. 1956. A New Moist Chamber Respirometer for the Study of the Metabolism of Tissue Slices in Vitro. J. Biol. Chem., 221, 71–7.
4. Pardee, A. B., Heidelberger, C., and Potter, V. R. 1950. The Oxidation of Acetate-1-C^{14} by Rat Tissue in Vitro. J. Biol. Chem., 186, 625–35.
5. DeDuve, Chr., and Berthet, J. 1954. The Use of Differential Centrifugation in the Study of Tissue Enzymes. Intern. Rev. Cytol., 3, 225–75.
6. Lehninger, A. L. 1955. Fatty Acid Oxidation in Mitochondria. In Methods in Enzymology, I, 545–8. S. P. Colowick and N. O. Kaplan, editors. Academic Press. New York.
7. Hunter, F. E., Jr. 1955. Coupling of Phosphorylation with Oxidation. In Methods in Enzymology, II, 610–6. S. P. Colowick and N. O. Kaplan, editors. Academic Press. New York.

EXPERIMENT 27. ESTERASE ACTIVITY OF RAT LIVER HOMOGENATE (1 period)

OBJECTIVES

In this experiment the activity of an esterase will be studied, as a function of substrate concentration. Experience will be gained in preparation of a homogenate, gassing of Warburg flasks, and manometric measurement of acid production.

PRINCIPLES

Enzyme biochemistry in reality includes two quite dissimilar areas. In the first we ask questions about enzymes as chemicals — their physical properties, chemical composition, catalytic activities as a function of experimental conditions, and how their structures enable them to act as catalysts. The second area concerns the discovery of the chemical reactions catalyzed by the enzymes and how these reactions are related in living organisms. These distinct problems are best studied in different types of tissue preparations — the first with pure enzymes and the second with enzymes as nearly as possible in their natural state. In general, the more closely a tissue preparation approaches the in vivo state, the less likely it is to be useful for the study of details of reactions.

The homogenate is among the most useful of a variety of tissue preparations that permit study of single reactions under more or less "natural" conditions (1, 2). In a homogenate, ideally, all cells are disrupted in a gentle manner and their contents are dispersed through the suspending medium. As a result, substrates and cofactors are diluted, and many reactions cannot proceed unless these components are added to the medium. We are able therefore to measure a single reaction of a metabolic sequence by adding the cofactors and substrate for this one reaction alone. Although homogenates bridge the gap between whole cells and purified enzymes to some extent, extrapolation in either direction must be made with caution.

In this experiment a homogenate of rat liver will be the enzyme source for the hydrolysis of an ester (3):

$$R-COOR' + H_2O \longrightarrow R-COO^- + R'OH + H^+$$

The Warburg respirometer can be used to follow this and other reactions which produce H^+, provided that the flask contains an equilibrium mixture of H_2CO_3 and HCO_3^-.

$$H^+ + HCO_3^- \rightleftharpoons H_2CO_3 \rightleftharpoons H_2O + CO_2$$

Any H^+ produced reacts with HCO_3^- to form gaseous CO_2 and the pressure change thus provides a measure of the amount of acid produced, as long as HCO_3^- is available. However, one mole of H^+ releases less than one mole of CO_2 because some of the H^+ is neutralized by other buffers in the reaction system. The proteins, for example, may combine with considerable H^+ (and may, in fact, control the pH of the solution). Therefore, one cannot use the flask constant calculated for CO_2 to determine the amount of H^+ produced, because of this "retention" of H^+. The relation between acid produced and CO_2 evolved must be determined experimentally by addition from the side-arm of an accurately known amount of acid to the complete reaction mixture and measurement of the resulting pressure change. This measurement need not be made for all flasks which contain the same buffers (and therefore have the same retention) because the required flask constant for each flask will be proportional by the same factor to the one calculated from calibration (4).

PRINCIPAL EQUIPMENT AND SUPPLIES

Warburg respirometer
10 Warburg flasks with side-arms
Potter-Elvehjem homogenizer (2)
Watch or stop watch
One rat (200 – 300 g) or mouse (15 – 25 g)
0.1 $\underline{M}$ $NaHCO_3$ (8.4 g $NaHCO_3$ per liter)
5% triacetin in water
Cylinder of 95% N_2 – 5% CO_2
Cold isotonic (0.15 $\underline{M}$) KCl
0.050 $\underline{N}$ H_2SO_4

PROCEDURE

Pipet into each flask 0.5 ml of 0.1 $\underline{M}$ $NaHCO_3$, variable amounts (0 to 2.1 ml) of 5% triacetin, and water to make a total of 2.6 ml; also pipet 0.2 ml of 0.050 $\underline{N}$ H_2SO_4 into the side-arm. Place the flasks in ice.

Kill and bleed the rat or mouse as in Experiment 26, and remove the liver. Weigh about 0.2 g of liver and homogenize it in a Potter-Elvehjem homogenizer (2) with 5 ml of cold isotonic KCl for about 2 min until particles are no longer visible. Dilute the rat liver homogenate to 10 mg of wet liver per ml (mouse liver: 5 mg/ml) and pipet 0.2 ml into each chilled, prepared Warburg flask at once. Boiled homogenate with 2.1 ml triacetin and 0.5 ml 0.1 $\underline{M}$ $NaHCO_3$ should be used as a blank. Attach the flasks to the manometers, make sure the side-arm vents are open, and pass the N_2–CO_2 mixture in through the manometer three-way stopcock and out through the side-arm for 5 min (5). Watch the manometer fluids closely, and adjust the pressure to prevent the fluid from being blown out. Transfer the flasks and manometers to the water bath, set the manometer fluid in the open arm near 0 and after 10 min take readings at 5 min intervals for 25 min. Then add the side-arm contents and make three more readings. Determine the approximate final pH of each flask with a drop of phenol red or bromthymol blue.

TREATMENT OF DATA

1. Plot the pressure change per 5 min for each flask against time. Measure the pressure change brought about by the added acid, and from

this value calculate the μ moles of acid equivalent to 1 mm of Brodie's solution pressure change.

2. Determine the initial rate of acid production for each flask.

3. Plot this initial rate against triacetin concentration in order to obtain the Michaelis-Menten constant and k_3E (see Experiment 20).

QUESTIONS

1. What particles are present in a homogenate? What functions are these particles thought to have in the cell? How are they isolated?

2. Does the osmotic pressure or ionic strength of medium in which a homogenate is prepared affect the enzyme activities? Why?

3. Compare homogenates and acetone powders as materials for metabolic study and as starting materials for enzyme isolation.

4. What would be the effect on your results of 0.1 M phosphate buffer, pH 7, in the flasks? What would be the effect of doubling the HCO_3^- concentration?

5. A homogenate catalyzes the reaction

$$A + H_2O \longrightarrow B^- + C + H^+$$

The pH optimum is 8.0. A absorbs light strongly at 280 mμ and B^- absorbs rather weakly at 420 mμ. Briefly compare three or more assay methods in terms of their utility for kinetic studies or as aids to purification of the enzyme. Stress sensitivity and convenience.

ALTERNATIVE EXPERIMENT, 27

Dehydrogenase activity may be studied with MnO_2 as a terminal hydrogen acceptor (6). Measurement depends upon uptake of CO_2 by combination with $Mn(OH)_2$ formed in the reaction.

References

1. Umbreit, W. W., Burris, R. H., and Stauffer, J. F. 1957. Manometric Techniques, 3rd ed., Chap. 10. Burgess Publishing Co. Minneapolis, Minn.

2. Umbreit, W. W., Burris, R. H., and Stauffer, J. F. 1957. Manometric Techniques, 3rd ed., Chap. 9. Burgess Publishing Co. Minneapolis, Minn.

3. Conners, W. M., Pihl, A., Dounce, A. L., and Stotz, E. 1950. Purification of Liver Esterase. J. Biol. Chem., 184, 29–36.

4. Umbreit, W. W., Burris, R. H., and Stauffer, J. F. 1957. Manometric Techniques, 3rd ed., Chaps 3 and 7. Burgess Publishing Co. Minneapolis, Minn.

5. Umbreit, W. W., Burris, R. H., and Stauffer, J. F. 1957. Manometric Techniques, 3rd ed., Chap. 5. Burgess Publishing Co. Minneapolis, Minn.

6. Hochster, R. M., and Quastel, J. H. 1952. Manganese Dioxide as a Terminal Hydrogen Acceptor in the Study of Respiratory Systems. Arch. Biochem. and Biophys., 36, 132–46.

EXPERIMENT 28. DEMONSTRATION OF A METABOLIC STEP WITH BACTERIAL MUTANTS (2 periods)

OBJECTIVES

This experiment illustrates the use of mutants to demonstrate metabolic pathways. Experience will be acquired in culture of microorganisms and in isolation and identification of a metabolic intermediate.

PRINCIPLES

Intermediary metabolism is a subject to which great effort is devoted in present-day biochemistry. It is clear that the syntheses of even the most complex metabolic intermediates, of molecular weights as much as 1000, occur in a sequence of simple reactions such as hydrations, dehydrogenations, decarboxylations, etc.; and it is the aim of intermediary metabolism to discover these individual steps and to assess their quantitative importance in the cell. Three experimental techniques have been most valuable in obtaining this information (1). First, radioactive compounds are used to show conversion of all or a portion of one molecule to all or part of another (see Experiments 36 and 37). A second method involves a direct demonstration of catalysis of the reaction under study by a cell-free preparation (see Experiments 29 and 30). Experiments of this sort are designed to demonstrate conversion of one compound into another; the enzymes are rarely isolated. The third method makes use of mutant organisms which cannot carry out a specific reaction; such mutants will be used in the present experiment. Normal cells, in the presence of enzyme inhibitors, sometimes may provide data of the same sort as do mutants.

There are two techniques for the study of metabolic pathways with mutants: intermediate accumulation or precursor utilization. Suppose that a mutant is unable to perform the reaction $F \rightarrow G$, where F, G, etc., are compounds of a metabolic sequence $A \rightarrow B \rightarrow \rightarrow K$. Then, after A is fed to the organism, the intermediate F may be accumulated inside the cell or in the medium, and one may be able to identify it. Alternatively, G or some other precursor of the final product, K, may be converted to K and may permit growth of the mutant; thus it will be implicated in the metabolism of K.

The present experiment utilizes the following sequence of reactions for pyrimidine biosynthesis in Escherichia coli (2, 3): aspartate + carbamyl phosphate $\rightarrow$ ureidosuccinic acid $\rightarrow$ dihydroorotic acid $\rightarrow$ orotic acid $\rightarrow$ orotidylic acid $\rightarrow$ uridylic acid $\rightarrow \rightarrow$ nucleic acids. A mutant which lacks the enzyme for the third reaction should (ideally) accumulate dihydroorotic acid and should be able to grow on orotic acid, orotidylic acid, or uridylic acid. If these events could be demonstrated experimentally, they would implicate these four compounds in pyrimidine biosynthesis. Interpretation of results obtained by these methods is uncertain for a number of reasons (4); however, valuable clues to pathways are often obtained with mutants.

PRINCIPAL EQUIPMENT AND SUPPLIES

  Autoclave
  Colorimeter
  Spectrophotometer
  Water bath or incubator
  Centrifuge
  Escherichia coli mutants 63–86 (ATCC 12632) and 550–460 (ATCC 11548)
  Uracil
  Orotic acid

PROCEDURE

1. *Preparation of Media*

   The basal medium contains 14 g of K_2HPO_4, 4 g of KH_2PO_4, 1.2 g of Na_3citrate·$5H_2O$, 0.2 g of $MgSO_4$·$7H_2O$, 2 g of $(NH_4)_2SO_4$, 10 ml of glycerol, and 2 liters of distilled water.

   a. Dissolve 16 mg of uracil in 1600 ml of basal medium. Put 100 ml into each of two 500 ml flasks, 70 ml in a 300 ml flask, and distribute the remainder (about 1300 ml) equally among four 1 liter flasks. The 1 liter flasks are to be used in the second part of this experiment.

   b. Prepare 70 ml of a solution that contains 10 μg/ml of orotic acid dissolved in basal medium.

   c. Place 300 ml of basal medium in a 1 liter flask.

   d. Autoclave all of the flasks.

2. *Effects of Uracil or Orotic Acid on Mutant Growth and Production of Intermediates*

   Four hours before you wish to harvest the bacteria, inoculate one 500 ml flask of basal medium plus uracil with 10^{10} cells of a stock culture of mutant 63–86 (grown overnight on basal medium plus 30 μg/ml uracil), and the other 500 ml flask with the same number of cells of mutant 550–460. The new suspensions should give a reading of 10 with the Klett colorimeter and filter No. 54 (540 mμ). Aerate the cultures at 37° by shaking or bubbling air through them until a density of 10^9 bacteria/ml is reached (turbidity with the Klett colorimeter equal to 100). Centrifuge at 6000 rpm for 5 min. Resuspend each mass of packed cells in 100 ml of basal medium. Put 20 ml of the suspension of mutant 63–86 into each of three 300 ml flasks containing 30 ml of basal medium, basal medium plus 10 μg/ml of uracil, and basal medium plus 10 μg/ml of orotic acid, respectively. Distribute the other mutant in the same way in three other flasks. Aerate the six flasks at 37°, and, for two hours, follow growth by reading the turbidity of aliquots at half-hour intervals. At 1 hr centrifuge 5 ml of each of the six cultures and read the optical densities of the supernatant fluids between 250 and 300 mμ at 10 mμ intervals in the spectrophotometer. Also determine spectra of the three original media, appropriately diluted.

3. *Isolation and Identification of Orotic Acid*

Twenty-four hr before performing Experiment 29, inoculate the remaining 1300 ml of basal medium plus uracil with three drops per flask of a dense culture of mutant 550−460, using sterile technique. Aerate the culture at 37°, preferably by shaking; otherwise bubble air through water first to saturate it before passing it through the flasks. Remove the bacteria by centrifugation and store them at 0° for Experiment 29. Dilute 0.3 ml of the supernatant solution with 2.7 ml of water and read the optical density at 290 mμ; the reading should be about 0.3. Assuming that 80% of this absorption is due to orotic acid ($\epsilon_{290} = 6.2 \times 10^3$; mol wt of orotic acid monohydrate = 174), how much orotic acid is in the solution? Concentrate the medium to about 80 ml by gentle boiling while an air stream blows onto the surface. Add a little charcoal to remove most of the yellow color and filter the solution while hot. After the solution is chilled overnight, potassium orotate should crystallize out. Separate the precipitate by centrifugation, dissolve most of it in as small a volume of boiling water as possible (less than 20 ml), add charcoal if much color is still present, and filter while hot. Bring the hot filtrate to 0.1 N with HCl and allow orotic acid to crystallize by cooling the solution to 0° overnight. Filter, wash the crystals with cold water, and dry them in vacuo.

Identify the crystals as orotic acid monohydrate by some of the following criteria: (a) melting point 323° (decomposes); (b) spectra at several pH (5); (c) equivalent weight and pK_a' by titration; (d) paper chromatography. For paper chromatography place 5 μl of 0.1% orotic acid neutralized with NaOH on Whatman No. 1 paper, develop the descending chromatogram with a solution containing 50 ml of n-butanol, 15.8 ml of 95% ethanol, 11.4 ml of 88% formic acid, and 22.8 ml of H_2O. Orotic acid gives a spot with an $R_f = 0.43$ that can be seen under ultraviolet light. Run a sample of authentic orotic acid for comparison.

TREATMENT OF DATA

Include in your notebook a table of results obtained in part 2 of this experiment. Indicate what compound(s) each mutant appeared to produce, what compound(s) permitted growth of each mutant, where each mutant was blocked, and what conclusions you can draw from your data regarding the role of orotic acid in pyrimidine biosynthesis. Suggest alternative explanations of your data.

QUESTIONS

1. Write the structural formulas of all compounds involved in this experiment. Which compounds have optical isomers? Which compounds have tautomeric forms?

2. Make a hypothesis to explain the observations given below and critically discuss the data in relation to it. What decisive experiments could be performed to check your hypothesis?

a. Wild-type (non-mutant) bacteria grow on X but do not accumulate compound L; nor do they use L even after growth on X in the presence of L.

b. A mutant strain requires compound $\underline{F}$ for growth and accumulates $\underline{L}$ when fed $\underline{X}$.

c. A tenfold purified enzyme preparation from the wild-type (nonmutant) organism converts $\underline{L}$ to $\underline{Q}$, but does so slowly, relative to the rate that would be required as a step in the production of $\underline{F}$ for growth.

d. Upon addition of $\underline{X}$-C^{14} and excess $\underline{Q}$, the wild-type bacteria accumulate $\underline{Q}$-C^{14}.

References

1. Davis, B. D. 1955. Intermediates in Amino Acid Biosynthesis. Advances in Enzymology, 16, 247−312. Interscience Publishers. New York.
2. Lieberman, I., and Kornberg, A. 1954. Enzymatic Synthesis and Breakdown of a Pyrimidine, Orotic Acid. J. Biol. Chem., 207, 911−24.
3. Yates, R. A., and Pardee, A. B. 1956. Pyrimidine Biosynthesis in Escherichia coli. J. Biol. Chem., 221, 743−56.
4. Adelberg, E. A. 1953. The Use of Metabolically Blocked Organisms for the Analysis of Biosynthetic Pathways. Bacteriol. Rev., 17, 253−67.
5. Shugar, D., and Fox, J. J. 1952. Spectrophotometric Studies of Nucleic Acid Derivatives and Related Compounds as a Function of pH. Biochim. et Biophys. Acta, 9, 199−218.

EXPERIMENT 29. ENZYMATIC OXIDATION OF DIHYDRO-
OROTIC ACID (1 or 2 periods)

OBJECTIVES

The demonstration of enzymatically catalyzed reactions in vitro as
evidence for metabolic pathways in vivo is illustrated in this experiment.
The metabolic step linking dihydroorotic acid and orotic acid will be
studied.

PRINCIPLES

The most conclusive proof that a particular reaction occurs in some
tissue is the demonstration of catalysis of the reaction by a cell-free
preparation of that tissue. Even with evidence of this type it is desirable
to obtain additional evidence by mutant or radioactive tracer techniques
in order to be certain that the reaction is important in the intact cell (1).

The methods that must be used to establish activity of a new enzyme
in vitro are basically the same as methods used previously for other enzymes; yet the task is not a simple one because of the large number of
variable conditions, such as choice of proper substrates, cofactors, pH,
and method of preparation of the enzyme, that may be involved. Essentially, one must discover conditions under which the desired enzyme is
active in vitro and then develop a reliable measure of activity. It is highly
instructive to read original articles in which studies of this sort have
been presented (see, for example, recent issues of Journals devoted to
biochemistry, and refs. [1, 2, 3]).

The initial step, which consists of discovering some evidence of enzyme activity in a cell-free extract, is the most difficult. A hypothesis is
made as to what the substrates of the enzyme are, based on prior observations of one's own or from the literature, on analogy with other systems, or on structural similarities (4). Next, a method is devised for de-
tecting a change in one of the suspected substrates—preferably a sensitive
and specific change such as can be measured by a colorimetric test or by
use of radioactive substrates to obtain radioactive products that are easy
to detect upon separation from the substrate. The extract and substrates
are incubated together, but it is not likely that activity will be observed
on the first attempt. Then one must search with the hope of finding a set
of conditions that give a demonstrable reaction. Some possible aids in
this search are these:

a. Use care in preparing the tissue extract (low temperatures, physio-
logical media), in order to prevent loss of enzyme activity.

b. Use a concentrated extract to avoid dilution of cofactors. However,
in a concentrated preparation there is danger of interference by side reactions, inhibitors, etc.

c. Add to the extract various coenzymes, Mg^{++}, Mn^{++}, other metal
ions, yeast extract, juice of boiled tissue (Kochsaft), etc.; try various
concentrations of these, as well as different pH.

d. Try different substrates, or derivatives of suspected substrates.

e. Try a different assay, such as one based on another reactant or anticipated product.

f. Prepare the extract from another organism or tissue.

The second stage in demonstration of an enzymatic reaction consists of systematic experimentation to improve the assay by finding optimal conditions and more convenient methods. It may be advantageous to purify the enzyme slightly (fivefold to twentyfold) in order to eliminate side reactions and to reduce assay blanks. To obtain an adequate description of the reaction it is often necessary to determine both the quantities of products that are formed and of reactants that disappear.

PRINCIPAL EQUIPMENT AND SUPPLIES

Beckman spectrophotometer and silica cuvettes
Centrifuge
Stop watch
Sonic oscillator (optional)
Thunberg tubes
Escherichia coli mutant 550− 460 (ATCC 11548) (or E. coli B)
Levigated alumina
100 Ml of 0.05 $\underline{M}$ phosphate buffer, pH 7.5
10 Ml of 0.02 $\underline{M}$ dihydroorotic acid (mol wt = 158) in 0.05 $\underline{M}$ phosphate buffer
Methylene blue or 2,6-dichloroindophenol (1 mg/ml in water)

PROCEDURE

1. *Preparation of the Extract*

Suspend the cells of mutant 550−460 which were harvested from 1300 ml of medium in Experiment 28 in cold, distilled water and wash them by centrifugation. Disrupt the bacteria either by suspending them in about 10 ml of buffer and treating them for 10 min at full power in the Raytheon 9 kc sonic oscillator; then remove the debris by centrifugation. Or, grind the wet pellet with 3 times its weight of "levigated alumina" for 3 min in a chilled mortar. Take up the ground material in 10 ml of buffer and remove the alumina and debris by low-speed centrifugation.

2. *Assay of Dihydroorotic Acid Dehydrogenase* (3, 5)

The conversion of dihydroorotic acid to orotic acid can easily be observed in the spectrophotometer since orotic acid strongly absorbs light at 290 mμ, while the absorption at this wavelength by dihydroorotic acid is negligible. Dilute a portion of the bacterial extract with buffer at 25° (about a hundredfold) to make a solution of optical density at 290 mμ about 0.7 units. Place 2.7 ml of this solution into each of 2 silica cuvettes. Mix the first with 0.3 ml of phosphate buffer and use it as a blank to adjust the spectrophotometer. Start the stop watch when you add 0.3 ml of 0.02 $\underline{M}$ dihydroorotic acid to the second sample. Read the optical density at 290 mμ

at 1 min intervals for 10 min. The initial rate of orotic acid production is taken as a measure of enzyme activity. Repeat this experiment at two other enzyme concentrations.

3. *Further Experiments* (as time permits)

a. Assay dihydroorotic dehydrogenase by the Thunberg method (6). Place 2 ml of phosphate buffer and 0.1 ml of 2,6-dichloroindophenol (or methylene blue) in the main parts, and 2.4 ml of various dilutions (about tenfold) of extract in the top parts, of sets of duplicate Thunberg tubes; then pipet 0.5 ml of 0.02 $\underline{M}$ dihydroorotic acid in the main part of one tube and 0.5 ml of buffer in the other of each pair. Evacuate the tubes, equilibrate them in a constant temperature bath, mix the contents, and determine the time required to reduce the blue color to the level of a standard that contains 0.5 ml of fiftyfold diluted dye in 4.5 ml of buffer. A more recent modification, in which tetrazolium compounds are reduced to purple-colored formazans may be tried (7).

b. Attempt a step in the purification of the enzyme, such as adsorption and elution from a gel or starch electrophoresis (see Experiment 19 for general references).

c. Devise experiments to test the effects of possible inhibitors, or make kinetic studies, as in Experiment 20.

TREATMENT OF DATA

1. Plot your data as μ moles of orotic acid formed vs. time.
2. Summarize your data for the results of part 3.

QUESTIONS

1. What conclusions do you draw from the combined results of Experiments 28 and 29 regarding the role of orotic acid in pyrimidine metabolism? Examine your conclusions critically. What would you do next if this were a problem in research?

2. Would you expect to find dihydroorotic acid dehydrogenase in mutant 63 – 86 (see Experiment 28)? Give your reasons for or against this conclusion.

3. Are your results consistent with the idea that the nucleotides of ureidosuccinic acid or dihydroorotic acid are not normal biosynthetic intermediates?

4. Outline an experiment with radioactive compounds that would provide information on pyrimidine biosynthesis. What would this experiment tell you? (See ref. [8].)

References

1. Davis, B. D. 1955. Intermediates in Amino Acid Biosynthesis. Advances in Enzymology, 16, 247–312. Interscience Publishers. New York.

2. McElroy, W. D., and Glass, H. B., editors. 1955. A Symposium on Amino Acid Metabolism. Johns Hopkins Press. Baltimore, Md.

3. Lieberman, I., and Kornberg, A. 1954. Enzymatic Synthesis and Breakdown of a Pyrimidine, Orotic Acid. J. Biol. Chem., 207, 911−24.

4. Bloch, K. 1956. Chemical Structure as a Guide to the Study of Biochemical Syntheses. In Currents in Biochemical Research 1956, 474−492. D. E. Green, editor. Interscience Publishers. New York.

5. Yates, R. A., and Pardee, A. B. 1956. Pyrimidine Biosynthesis in Escherichia coli. J. Biol. Chem., 221, 743−56.

6. Umbreit, W. W., Burris, R. H., and Stauffer, J. F. 1957. Manometric Techniques, 3rd ed., Chap. 8. Burgess Publishing Co. Minneapolis, Minn.

7. Mii, S., and Green, D. E. 1954. Reconstitution of Fatty Acid Oxidizing System with Triphenyltetrazolium as Electron Acceptor. Biochem. et Biophys. Acta, 13, 425−32.

8. Roberts, R. B., Abelson, P. H., Cowie, D. B., Bolton, E. T., and Britten, R. J. 1955. Studies on Biosynthesis in Escherichia coli. Carnegie Institution of Washington Publication 607. Washington, D.C.

EXPERIMENT 30.   GLUTAMINE SYNTHESIS (1 or 2 periods)

OBJECTIVES

This experiment provides an example of a biosynthetic reaction in which energy from ATP is utilized. The techniques for preparation of an acetone powder and of colorimetric assay for enzyme activity are illustrated. Activation by metal ions and competitive or non-competitive enzyme inhibition may also be studied.

PRINCIPLES

The enzyme glutamine synthetase (1, 2) catalyzes the reaction

glutamic acid + ATP + ammonia $\longrightarrow$ glutamine + ADP + phosphate

To make the assay easier, the ammonia in the reaction mixture may be replaced by hydroxylamine (NH_2OH), with the consequent formation of glutamyl hydroxamic acid. The hydroxamic acid is determined colorimetrically as its Fe^{+++} complex.

Glutamine synthesis is representative of a large group of reactions that require ATP as a source of energy. The mechanism of glutamine synthesis is not yet completely known and active research on this problem is now in progress (1).

PRINCIPAL EQUIPMENT AND SUPPLIES

Colorimeter
20 G dry green-pea seeds (not treated with fungicide)
200 Ml 0.1 $\underline{M}$ NaHCO$_3$ (cold)
1 M MgSO$_4$
300 Ml of acetone at $-10°$
10 Ml 0.025 $\underline{M}$ ATP, pH 7.5 (16 mg/ml of Na$_2$H$_2$ATP·4H$_2$O)
0.5 $\underline{M}$ sodium glutamate, pH 7.5
0.8 $\overline{M}$ tris(hydroxymethyl)aminomethane buffer, pH 7.5
10 $\overline{Ml}$ 1 M NH$_2$OH adjusted to pH 7.5. Prepare a fresh solution from NH$_2$OH·HCl just before use.

Ferric chloride reagent: equal volumes of 10% FeCl$_3$·6H$_2$O in 0.2 $\underline{N}$ HCl, 24% TCA, and 50% (v/v) HCl are mixed together.

PROCEDURE

1. *Preparation of Enzyme Extract*

Soak 18 g of dried peas overnight in water. Grind them in a chilled mortar with addition of small portions of cold 0.1 $\underline{M}$ NaHCO$_3$ until a total of 20 ml has been added. Add 2 ml 1 $\underline{M}$ MgSO$_4$ and store the preparation at 0° for 30 min. Centrifuge the suspension and discard the debris. The supernatant solution may be used directly as a source of enzyme. Alternatively, an acetone powder of the peas may be made (3) as follows: Chill the soaked ground peas to $0°$ and then stir them into 200 ml of acetone at $-10°$. Remove the acetone by vacuum filtration; rinse the residue with more cold acetone and then with ether. Avoid drawing air through the preparation. Spread the residue on a paper towel and allow it to dry in air or under vacuum. Store the powder in a cold box $(-20°)$. The acetone powder is extracted with 5 ml of cold $0.1 \underline{M}$ NaHCO$_3$ per gram of powder, and insolubles are removed by centrifugation. The solution is an active enzyme preparation.

2. *Assay of Glutamine Synthetase* (2)

The assays are performed in 12 ml centrifuge tubes. The complete mixture contains 0.5 ml of buffer, 0.5 ml of ATP, 0.5 ml of glutamate, 0.1 ml of $1 M$ MgSO$_4$, 0.1 ml of NH$_2$OH, variable amounts (up to 0.5 ml) of enzyme, and water to a total of 2.2 ml. After incubation for 20 min at $30°$ add 0.75 ml of ferric chloride reagent and remove the precipitated material by centrifugation. Determine the hydroxamic acid concentration in the supernatant solution by measurement in the spectrophotometer at 540 mμ within 30 min. Determine the activity at three different levels of enzyme concentration and include appropriate blanks.

3. *Activation and Inhibition of Glutamine Synthetase*

Devise an experiment to demonstrate activation by various concentrations of Mg^{++}; or demonstrate competitive inhibition by methionine sulfoxide or ADP, or non-competitive inhibition by F$^-$ or Hg^{++} (2).

TREATMENT OF DATA

Calculate the units of synthetase activity in your preparation (2).

Plot your data for activation or inhibition as graphs of activity vs. concentration of activator or inhibitor. Determine the affinity constant of activator or inhibitor for synthetase if your data permit (4).

QUESTIONS

1. What advantages has an acetone powder, relative to a crude extract, for an experiment such as this one?

2. Can you suggest a mechanism whereby energy from ATP can be utilized to form the amide bond of glutamine?

3. What is the difference between competitive and non-competitive inhibition in terms of a supposed site of binding and in terms of mathematical formulation?

4. Discuss possible mechanisms of activation or inhibition of the enzyme by the substances listed in part 3 of the Procedure.

References

1. Meister, A. 1956. Metabolism of Glutamine. Physiol. Revs., 36, 103—27.
2. Elliott, W. H. 1955. Glutamine Synthesis. In Methods in Enzymology, II, 337—42. S. P. Colowick and N. O. Kaplan, editors. Academic Press. New York.
3. Nason, A. 1955. Extraction of Enzymes from Higher Plants. In Methods in Enzymology, I, 62—3 (see also 33—5, 55, 609—10). S. P. Colowick and N. O. Kaplan, editors. Academic Press. New York.
4. Alberty, R. A. 1956. Enzyme Kinetics. Advances in Enzymology, 17, 1—64. Interscience Publishers. New York.

RADIOACTIVE ISOTOPE TRACER TECHNIQUES IN BIOCHEMICAL RESEARCH

A. Procedures for Measurement of Radioactivity

OBJECTIVE

These experiments should provide an introduction to the techniques for measurement of radioactivity that are essential for proper execution of biochemical experiments with radioactive tracers.

PRINCIPLES (1 – 5)

Radioactive isotopes are unstable atomic species which disintegrate at rates which are expressed by the half-life of the isotope. In the process of disintegration subatomic particles (α, β, and γ) of high energy content are emitted. These particles cause ionization of molecules in their paths, and this phenomenon is exploited in the measurement of radioactive isotopes.

Figure 15 and the following explanation present a highly simplified description of the Geiger counter, which is one type of instrument for the detection and counting of particles from the disintegration of radioactive isotopes.* The Geiger tube contains two electrodes across which a poten-

---

* For the sake of brevity and simplicity, only the end-window Geiger counter will be described and used in the experiments of this Section. Other types of instruments are described in specialized texts, such as ref. (2).

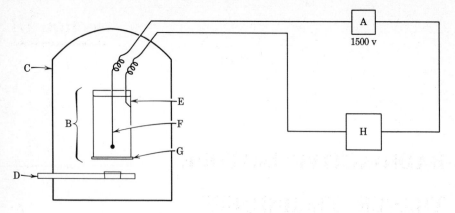

Figure 15. Schematic diagram of a Geiger–Müller tube and scaler.

A. Source of potential   E. Cathode sheath
B. Geiger–Müller tube   F. Anode wire
C. Lead shield   G. Mica window (1.5–2.0 mg/cm^2)
D. Sample pan holder   H. Scaler

tial difference of 1200–1500 volts is maintained. The tube is filled with
helium at low pressure plus a small amount of other, so-called "quenching" gases. One end of the tube is sealed by a thin mica window beneath
which the radioactive sample is placed, and both the tube and the sample
are held within a heavy lead housing to reduce extraneous or "background" radiations from such sources as cosmic rays. A particle of radiation (α, β, or γ) from the sample penetrates the window and causes
ionization of gas molecules and concomitant release of electrons. The
free electrons are accelerated toward the positive terminal of the tube.
These electrons strike and ionize other gas molecules in their path, until
the result is a brief, measurable surge of current between the two electrodes. This impulse of current is detected by the electronic circuit of
the "scaler" and recorded on a mechanical register as one count.

It is important to supply the correct electrical potential across the
electrodes of the Geiger tube (Experiment 31). As the difference in potential between the terminals is increased, a threshold voltage is reached
at which ionizing particles begin to be detected (Figure 16). The fraction
of ionizing particles detected then increases with increasing tube voltage
until a plateau is reached. Within the plateau, the efficiency of the tube is
nearly independent of the potential across it. This is the so-called
"Geiger-Müller region," and most radioactive samples are counted in
this region. Care should be taken not to exceed the upper plateau voltage,
for the tube is easily damaged at higher voltages.

Not all the particles that leave the sample will strike and penetrate
the Geiger tube window. Some of the particles will be directed toward the
counter window; others will be directed down, away from the counter window. A fraction of the latter will be deflected by the sample and the sample

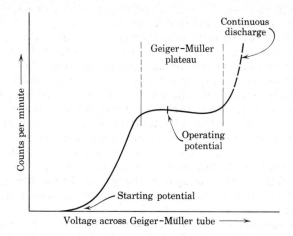

Figure 16. Sensitivity of the Geiger-Müller tube at different voltages.

pan and directed toward the window. This is called <u>backscatter,</u> and the fraction of the radiation that is backscattered depends upon the type of radiation, nature and size of sample, and composition and dimensions of sample pan. If the nature of the sample, type of sample pan, and geometry (i.e., the area of sample, distance from counter window, etc.) are kept constant from sample to sample, one will measure a constant fraction of the total activity, and the relative activities of the various samples will be comparable. From the number of counts recorded for a sample, there must be subtracted the so-called <u>background count</u>, which is the number of counts/min recorded in the absence of a sample. The background is the result of cosmic rays that penetrate the lead housing or of radiation from contaminants in the counting area.

The detection of a given high-energy particle by the counting system requires a finite time interval during which no other particles can be detected. This interval is called the resolving time of the system. The interval is slightly different for various types of apparatus, but it is usually of the order of 5 microseconds. When samples of comparatively high activity are counted, the resolving time of the tube limits the proportion of the total counts detected, and a correction factor must be applied. This <u>coincidence correction</u> should be determined with the equipment and under the same conditions that will be used for experimental samples (see Experiment 34). Samples of less than 2500 ct/min need not be corrected for coincidence if a counter of the general type described in this manual is employed.

The efficiency of the counting system may vary slightly from day to day because of slight variations in line voltage, condition of electronic tubes or counting tube, slight changes in geometry, or indeterminate factors. These changes are outside the normal statistical variations that result from the random nature of the decay process. To correct for this variable efficiency, it is necessary to count a standard sample with a known and fixed number of disintegrations per minute on each day that unknown samples are counted (Experiment 32).

The statistical error caused by the random nature of the decay process is inversely proportional to the square root of the total count; hence, the percentage error is decreased by increasing the count. The total count may be increased by increasing the size of the sample or the time during which the sample is counted. Since errors in corrections, such as self-absorption and coincidence corrections, become large as the sample size increases, an increase in the counting time is usually the best procedure. Counts should be expressed with a standard deviation or other statistical value that defines the range about the true count within which any given count may be expected to fall with a certain probability. For accurate results the counting time for sample and background should be so adjusted that the standard error for the net activity of the sample is only 1% or 2% (Experiment 33).

Certain properties of C^{14}, the radioisotope used in the next few experiments, deserve mention. The half-life of C^{14} is 5100 ± 200 years. For practical purposes, the activity of a given sample will not diminish during the lifespan of the investigator, and decay corrections are unnecessary. C^{14} emits a beta particle of low energy (0.15 million electron volts [MEV]). In order that an appreciable fraction of the total radiation of such a sample may penetrate the tube, it is necessary to have an extremely thin mica window (1-2 mg/sq cm) and to place the sample very close to the counter window. With these measures, one may detect 5% to 10% of the radiation produced by the C^{14} disintegration. A fraction of the emitted beta particles is absorbed in the substance of the sample itself, and the correction factor for self-absorption (Experiment 35) will vary with sample thickness and the nature of the sample material.

The various corrections of the observed radioactivity count rate must be applied in the proper sequence. The order for the application of these corrections is listed below.

a. The probable error is determined from the observed count.

b. The coincidence correction is applied to the observed count.

c. The background correction is applied to the count from b.

d. The self-absorption correction is applied to the count from c.

e. Any corrections for counter efficiency or sample geometry are applied to the count from d.

PRINCIPAL EQUIPMENT AND SUPPLIES

End-window, Geiger-Müller tube and scaler
Aluminum sample pans
Plate storage trays*
Flat, bent-tip forceps

---

*Plate storage trays may be made from 1 ft by 1 ft by 1 in. boards. Holes somewhat larger than the diameter of the sample pans are drilled to a depth of $\frac{3}{8}$ in. A hinged plastic cover completes the tray.

Sample spinner and dryer (Figure 17)
Filter apparatus (Figure 18)
Centrifuge
Infrared lamps
Orange shellac
Suitably diluted solutions of $Na_2C^{14}O_3$
C^{14}-labeled urea, sodium acetate, or other non-volatile, water-soluble compound
Barium chloride (0.5 $\underline{M}$ and 1.0 $\underline{M}$)
1.0 $\underline{M}$ NH_4Cl
Whatman No. 50 paper

PROCEDURE

Operation of Counters

An operation manual will be attached to the instrument. Read the manual thoroughly before you use the instrument and check up on points of uncertainty as they arise. If in doubt, ask the instructor; this is not the place for experimentation.

Some Observations on the Maintenance of an Uncontaminated Laboratory (2)

Reasonable care in the organization of the laboratory and instruction of the students should forestall accidents that might lead to permanent contamination. Areas for the counters and areas for the preparation of radioactive samples should be separated as far as possible. Areas for the preparation of radioactive samples should be clearly defined, and all tables in that area should be covered with wrapping paper. Separate containers for radioactive pipets, solid wastes, and liquid wastes should be provided; and the student should be informed of the level of radioactivity that may safely be washed down the sink. Any work in which radioactive gases, such as carbon dioxide, are released into the atmosphere should be conducted in a hood.

General Information for Preparation of Radioactive Plates*

Radioactive samples generally are counted most readily and accurately in the solid state. These samples are plated by evaporation from solution, or by evaporation or filtration from a slurry of insoluble material. Thin samples (less than 0.5 mg/cm^2) may be plated by evaporation from solution or by filtration from a slurry, but thin samples plated by evaporation from a slurry are irregular and unsatisfactory. A sample to be plated by evaporation from solution is usually applied from a pipet as a spiral onto the sample pan (Experiment 34), and one apparatus for such plating is shown in Figure 17. This technique is suitable for highly

---

*Chap. 7 of ref. (1).

active biological extracts. A convenient apparatus for plating by filtration is shown in Figure 18. The latter technique also is convenient for plating compounds that can be readily precipitated on the filter (see Experiment 35).

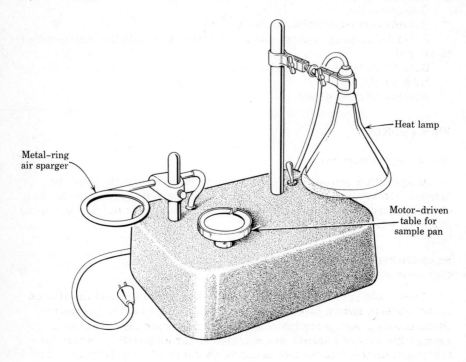

Figure 17. Sample spinner. (Courtesy of Nuclear Instrument and Chemical Corporation, Chicago, Ill.)

Thick samples also may be plated by evaporation from solution or by filtration from a slurry; in addition, thick samples may be plated by evaporation from a slurry (Experiment 32). Since self-absorption will vary with the nature of the sample, thick samples should be converted to $BaCO_3$ in order to obtain accurate counts. The organic material is burned to CO_2 and the CO_2 is absorbed in a solution of sodium hydroxide (Chap. 6 of ref. [1]). Upon addition of excess $BaCl_2$, the sample is converted to $BaCO_3$ which may be plated by the techniques mentioned above.

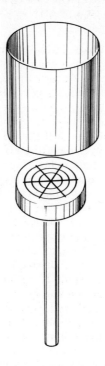

Figure 18. Apparatus for filtration plating. (Courtesy of Tracerlab, Inc., Boston, Mass.)

EXPERIMENT 31. DETERMINATION OF THE GEIGER-
MULLER PLATEAU* ($\frac{1}{2}$ period)

1. Assemble and turn on the counting equipment as described in the operation manual for that particular equipment.

2. Insert a radioactive sample under the counter tube, preferably a sample that yields at least 1000 counts per minute.

3. Extinguish the interpolation lights; reset the mechanical register and timer. Turn the stop-count switch to <u>count</u>.

4. Slowly increase the voltage until the threshold voltage is exceeded and counts begin to register on the scaling circuit.

5. Proceed from the threshold value. Increase the voltage in 50 volt steps, and take a one-minute count at each step in order to determine the counting rate as a function of the voltage (see Figure 16). Immediately beyond the threshold voltage a rapid rise in counting rate will be noticed until the Geiger plateau is reached. Do not increase the voltage beyond this plateau; the termination of the plateau will be noted by a second rapid increase in counting rate as the voltage is raised. If the voltage is increased beyond this point, a continuous discharge will result, and the Geiger-Müller tube will be damaged. Keep the stop-count switch in the count position while you change the voltage so that the continuous discharge region will be readily detected. The life of the tube may be considerably shortened by even a short period at the higher voltage.

6. Plot the counting rate per minute as ordinate against the voltage as abscissa.

7. Henceforth, operate your counter at a voltage within the first half of the Geiger-Müller plateau.

---

*Questions and references for Experiments 31−35 will be found on p. 149.

EXPERIMENT 32. PREPARATION OF RADIOACTIVITY
STANDARDS ($\frac{1}{2}$ period)

Standards for the calibration of counters may be obtained from the National Bureau of Standards. Secondary standards may be prepared readily in the laboratory, and these serve as a convenient index of the efficiency and precision of the counter, although they should not be relied upon for absolute disintegration rates. A convenient secondary standard for experiments with C^{14} is $BaC^{14}O_3$. It is advisable to have two standard plates, one yielding about 1000 counts per minute and another about 5000 counts per minute. These will reveal variations in coincidence correction as well as variations in counting rate at either of the levels.

Preparation of the Standard Plates

Calculate the amount of stock $Na_2C^{14}O_3$ (0.125 $\underline{M}$ Na_2CO_3; and 0.01 $\mu c/ml$*) required to yield plates of 1000 and 5000 counts/min. Consider the efficiency of the counter that you will use as well as the disintegration rate of the carbonate that you plan to plate; check your calculations with the instructor before proceeding.

Technique for plating by evaporation from a slurry

1. Pipet the required amount of stock $Na_2C^{14}O_3$ into a 12-ml centrifuge tube. CAUTION: use a rubber suction bulb for all pipetting of radioactive solutions.
2. For each 1.0 ml of carbonate, add 0.25 ml of 1 $\underline{M}$ NH_4Cl and 0.5 ml of 1 $\underline{M}$ $BaCl_2$.
3. Stir with a small stirring rod.
4. Centrifuge the solution and pour off the supernatant liquid.
5. Disperse the solid in a few ml of water.
6. Recentrifuge and pour off the supernatant liquid.
7. Repeat steps 5 and 6 twice, first with 50% ethanol and then with 95% ethanol.
8. Wash the solid onto a sample pan with 95% ethanol.
9. Dry the plate carefully under a heat lamp. (There should not be excess $BaCO_3$ around the edges of the dried plate; if there is, repeat the plating.) CAUTION: Keep radioactive pipets separated from the others, and mark them distinctly.

Preparation of permanent secondary standards. If a plate is required for use as a permanent secondary standard, some binding material should be added to the carbonate. The binder will prevent flaking and loss of activity through exchange of C^{14} of the carbonate with atmospheric carbon dioxide. A simple binder is one drop of diluted shellac added to the suspension for every 20 mg of barium carbonate. (Dilute 1 part of orange shellac with 4 parts of 95% ethanol.) Evaporation of the alcohol then is carried out in the usual manner, but the high-temperature drying is omitted. Such samples cannot be compared with samples plated without shellac; they are useful only as permanent secondary standards to reveal variations in the counter from day to day.

---

*1.0 μc = 1 × 10^{-6} curie (c); 1 curie is the amount of radioactive material that will yield a disintegration rate of 3.7 × 10^{10} disintegrations/second.

EXPERIMENT 33.   CALCULATION OF THE PROBABLE
ERROR IN COUNTING ($\frac{1}{2}$ period)

The object of this experiment is to count samples within a standard error of $\pm 1\%$. Obtain several samples from the instructor. First, insert the samples successively in the counter and determine the counting rate roughly for a one-minute interval. Determine the background similarly. From the approximate counting rate and background, calculate the length of time that both the background and samples must be counted to attain the statistical accuracy desired. Use the formulas given in ref. (1), 286–88. Count each sample for the calculated length of time. If C^{14} is used, no correction is necessary for decay, but all results should be corrected for background.

EXPERIMENT 34. DETERMINATION OF COINCIDENCE
CORRECTIONS ($\frac{1}{2}$ period)

An instructive way to determine the coincidence correction is to compare the measured count and the theoretical count for samples of increasing activity. For convenience, the samples should be of such specific activity that self-absorption may be neglected.

Prepare a solution of C^{14}-labeled urea (or sodium acetate,* or other economical source of non-volatile C^{14}) of 0.225 $\mu c/ml$ and containing about 1.0 micromole per milliliter of urea or other carrier compound. Calculate the expected count/min/ml from the counter efficiency and the $\mu c/ml$. Prepare six to eight plates that should give activities of approximately 25,000 ct/min down to 250 ct/min. Accurate delivery of volumes for plates of the lower activities will necessitate serial dilution of the original solution; the accuracy of the coincidence determination is limited by the accuracy with which these dilutions are made. Plates are conveniently formed by evaporation from solution.

Procedure for plating by evaporation from solution. The required volume of stock C^{14} solution is pipetted (with suction bulb) onto the sample plate in a gradually widening spiral while the plate is exposed to a current of air and to mild heat from a lamp. The area of the spiral should be less than the area of the counter window. A convenient apparatus is shown in Figure 17.

Count all the samples for sufficient periods to yield the desired statistical accuracy (see Experiment 33) and plot sample dilution vs. both the calculated ct/min and the observed ct/min after correction for background. The two curves should coincide at low activities (high dilutions), and the discrepancy between the two curves at higher activities constitutes the error due to coincidence. The coincidence correction factor is the ratio of the true count/observed count, and a given sample may be corrected for coincidence by multiplying the observed count by the coincidence correction factor for that count rate.

In most tracer work, the procedure outlined above is adequate since it is capable of a precision greater than 1%. Theoretical methods are of greater accuracy and these may be used if the counting system is known to follow equations relating the true counting rate, the apparent counting rate, and the resolving time. Details for these procedures are given in Appendix III of ref. (1).

---

*CAUTION: acetate is volatile when plated at pH values near or below the pK_a, or when it can form a volatile salt, such as ammonium acetate.

EXPERIMENT 35. RADIATION LOST BY SELF-ABSORPTION
(1 period)

A count for a given sample is corrected for self-absorption by multiplying the count by a self-absorption correction factor. The self-absorption correction is a function of sample thickness; the thickness may be expressed in mg/cm^2 or in mg per plate of a constant but not necessarily known area (see ref. [3], 89–93). To prepare such correction factors for $BaC^{14}O_3$ samples plated on filter paper, plate $BaC^{14}O_3$ samples of constant activity/mg of $BaCO_3$, but of increasing amounts of sample per plate. The stock $0.125\,\underline{M}\,Na_2C^{14}O_3$ solution will contain about 12,500 measureable ct/min/ml. Prepare samples of 0.5 to 12 mg as $BaCO_3$ per cm^2 of plate area by filtration plating.

Technique for plating carbonate by filtration

1. Wash sheets of Whatman No. 50 paper with water and then with 95% ethanol, dry them in air, and store them in a dessicator. Cut a disk of the desired size from this paper and weigh it.

2. Place the paper disk in the filtration apparatus shown in Figure 18 and moisten it with water. Draw off excess water with vacuum.

3. Disconnect the vacuum and then pipet into the filter cup 1.0 ml of $0.5\,\underline{M}\,BaCl_2$ for each 1.0 ml of stock $Na_2C^{14}O_3$ that is to be plated.

4. Quickly add the desired volume of the $Na_2C^{14}O_3$ solution and stir the mixture with a small, fire-polished rod.

(CAUTION: use a rubber suction bulb for pipetting the $Na_2C^{14}O_3$.)

5. Apply suction and wash the $BaC^{14}O_3$ with water, 95% ethanol, and acetone; do not allow the plate to dry until after the acetone wash. Disconnect the vacuum as soon as the plate whitens as a sign that excess acetone has been removed; do not continue the suction for an excessive time because it will cause cracked plates. Upon disassembly of the apparatus the filter pad should be flat, free of wrinkles, and covered with a uniform layer of $BaC^{14}O_3$.

6. Dry the plate in a vacuum oven at 50–60° or in a desiccator over $CaCl_2$.

7. Reweigh the plate to determine the total weight of $BaCO_3$.

8. Count each plate to what you consider to be the required statistical accuracy (see Experiment 33).

Plot the counts per minute vs. mg/cm^2 and draw a straight line with a slope equal to the initial slope of the curve. The straight line is equal to the theoretical rate at any thickness of sample. The self-absorption correction is the ratio of the apparent activity to the true activity at any thickness, and the accuracy of this ratio depends upon the accuracy with which the original slope was drawn. Tabulate the values of these ratios as percentages of the theoretical activity at various thicknesses.*

TREATMENT OF DATA

When Experiments 31 through 35 are complete, submit the data to the instructor in the form of clear, properly labeled charts and tables.

---

*It should be obvious that this data for self-absorption corrections is applicable only to $BaC^{14}O_3$ samples plated on filter paper, for self-absorption by $BaCO_3$, absorption by paper fibers, and backscattering from the sample and the filter paper are all factors that influence the observed count.

QUESTIONS

1. Describe at least two other types of instruments for radioisotope measurement that are based on principles other than those of the Geiger-Müller detector.

2. What is the purpose of the "quenching gas" in the Geiger-Müller tube?

3. Compare the half-life and particle emission of C^{14}, P^{32}, and I^{131}. How might the radioactivity of each be determined in a mixture of all three?

4. From your data for Experiment 35, determine what thickness of $BaCO_3$ would be required to have a sample of "infinite sample thickness." What are the advantages of plating to "infinite thickness"; what are the limitations?

References

1. Calvin, M., Heidelberger, C., Reid, J. C., Tolbert, B. M., and Yankwich, P. F. 1949. Isotopic Carbon. John Wiley and Sons. New York.

2. Comar, C. L. 1955. Radioisotopes in Biology and Agriculture. McGraw-Hill Book Co. New York.

3. Kamen, M. D. 1951. Radioactive Tracers in Biology, 2nd ed. Academic Press. New York.

4. Willard, H. H., Meritt, L. L., and Dean, J. A. 1951. Instrumental Methods of Analysis, 2nd ed. D. van Nostrand Co. New York.

5. Aronoff, S. 1956. Techniques of Radiobiochemistry. Iowa State College Press. Ames, Iowa.

B. Application of Radioactive Isotopes in Biochemical Research

The purpose of this group of experiments is to illustrate some of the general principles and techniques employed in the study of intermediary metabolism with tracers and to clarify certain calculations that arise in such research. Experiment 36 provides evidence with intact animals of the occurrence and interrelationship of glycolysis, fatty acid oxidation, and the tricarboxylic acid cycle (1); Experiment 37 reviews classical evidence for carbon dioxide fixation in propionic acid bacteria (2).

The experiments do not involve radioactivity levels that constitute a health hazard; however, carelessness could easily result in serious and permanent contamination of the laboratory. Carefully follow instructions on the disposal of radioactive wastes and respiratory carbon dioxide, the cleansing of contaminated glassware and instruments, and precautions against contamination of the laboratory by spillage of radioactive materials.

The authors recommend C^{14} for introductory experiments in radio-isotope technique because of its wide application in biochemical research, the minor health hazard, and the simplicity of the experiments. However, experiments with other isotopes have certain advantages that arise from their different radiation characteristics. For example, an experiment with inorganic phosphate (P^{32}) is instructive from the aspects of the shorter half-life ($T_{1/2} = 14.3$ days) and more powerful beta emission (1.72 MEV). A study of the rate of incorporation of P^{32} into yeast cultures grown under various conditions constitutes a simple experiment with this isotope. As an extension of this experiment, the P^{32}-labeled products in the yeast may be extracted and identified by techniques from Section I of this book. I^{131} is an isotope that is also instructive because of the short half-life $T_{1/2} = 8$ days) and powerful radiation. The authors have had success with an experiment for the incorporation of I^{131} in thyroid homogenates that is based on a paper by Fawcett and Kirkwood (3). The latter experiment also permits the student to become familiar with the technique of radioautography.

References

1. Lifson, N., Lorber, V., Sakami, W., and Wood, H. G. 1948. Incorporation of Acetate and Butyrate Carbon into Rat Liver Glycogen by Pathways Other than Carbon Dioxide Fixation. J. Biol. Chem., 176, 1263–84.
2. Wood, H. G., and Werkman, C. H. 1936. The Utilization of CO_2 in the Dissimilation of Glycerol by the Propionic Acid Bacteria. Biochem. J., 30, 48–53.
3. Fawcett, D. M., and Kirkwood, S. 1953. The Synthesis of Organically Bound Iodine by Cell-Free Preparations of Thyroid Tissue. J. Biol. Chem., 205, 795–802.

EXPERIMENT 36. METABOLISM OF C^{14}-LABELED FATTY ACIDS
IN THE INTACT RAT* (5 periods; students will
work as pairs throughout this experiment)

OBJECTIVES

This experiment will illustrate some of the general principles and
techniques employed in the study of intermediary metabolism with radioisotopes.

INTRODUCTION

In this experiment glucose and C^{14}-labeled fatty acids will be fed to
rats in which the liver glycogen has been depleted by a 24-hr fast. Part
of the fed material is oxidized and converted to respiratory CO_2. To determine the extent of oxidation of the fatty acid to CO_2, the rat will be
kept in a closed system, and the respiratory CO_2 will be collected in alkali.
Subsequently, the C^{14} in the respired CO_2 will be determined.

Another part of the fed material is converted to body constituents,
among others, liver glycogen. The extent of conversion of the fatty acid to
liver glycogen will be determined by isolation of the glycogen and measurement of the C^{14} content. In order to determine the distribution of C^{14}
in the glucose, the isolated glycogen will be hydrolyzed and the glucose
will be fermented by <u>Lactobacillus casei</u>. Fermentation by this organism
leads to a central cleavage of the glucose molecule into two molecules of
lactic acid. The lactic acid in turn will be oxidized with permanganate to
acetaldehyde and CO_2. It is possible to split the acetaldehyde into two
1-carbon fractions (CHI_3 and $HCOOH$) by the iodoform reaction, but that
will not be attempted in the present experiment. The degradation is illustrated below.

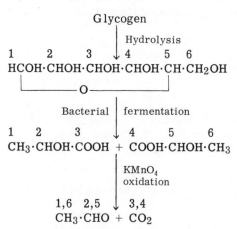

---

* This experiment was adapted from the laboratory manual of the Dept. of
Biochemistry, Western Reserve University, School of Medicine, and the experiment
is used with the permission of Western Reserve University, School of Medicine.

By this series of reactions carbons 3 and 4 of glucose will be obtained as CO_2 and the radioactivity of this fraction may be compared with that in the acetaldehyde which represents carbons 1, 2, 5, and 6 of the original glucose.

This information on the distribution of the C^{14} in the glucose carbon chain serves as an indicator of the path of intermediary metabolism. The distribution of the isotope can be predicted on the basis of a synthesis of glycogen via a given scheme, for example, glycolysis and the tricarboxylic acid cycle (1–4). If it is found that the C^{14} distribution is in agreement with the predictions of these latter paths, this evidence constitutes strong support for the hypothesis that these pathways apply in metabolism in vivo.

PRINCIPAL EQUIPMENT AND SUPPLIES

Period 1

> Rats of 200–250 g weight
> Cages
> Basic diet
> Balances

Period 2

Equipment as illustrated in Figure 19 (two sets for each pair of students)

> Centrifuge
> Syringes with special needles (see Figure 20) for injection of glucose and the fatty acids.
> C^{14}-labeled fatty acids as 1.25 M solutions in 25% glucose which contain 2 μc per millimole of fatty acid. Suitable acids are carboxyl- or methyl-labeled acetic or carboxyl-labeled propionic and butyric acids
> 30% KOH
> 1 N H_2SO_4
> 10% trichloracetic acid (should be made fresh for this experiment and stored in the refrigerator)

Period 3

> Standard glucose, 0.12 mg/ml
> Reagents A and B for glucose determination*
> Arsenomolybdate reagent†
> Tank of CO_2

---

* Reagent A: 25 g anhydrous Na_2CO_3 + 25 g KNa tartrate + 20 g $NaHCO_3$ + 200 g anhydrous Na_2SO_4 are dissolved in distilled water and diluted to 1 liter. Filter if necessary and store at 20° or higher. Any sediment that forms may be removed by filtration.

Reagent B: 15% $CuSO_4 \cdot 5H_2O$ that contains 1–2 drops of concd. H_2SO_4/100 ml.

† Dissolve 25 g of ammonium molybdate in 450 ml of distilled water and add 21 ml of concd. H_2SO_4. Dissolve 3 g of $Na_2HAsO_4 \cdot 7H_2O$ in 25 ml of water. Add the arsenate to the acid molybdate solution and mix well. Incubate the mixture at 37° for 1–2 days or warm it carefully at 55° for 25 min. Store in a brown bottle.

Suspension of Lactobacillus casei (50%)*
Water bath, preferably with a shaker platform
0.8 $\underline{M}$ NaHCO$_3$

Period 4

0.5% 2,4-dinitrophenylhydrazine in 2 $\underline{N}$ H$_2$SO$_4$†
Saturated Ba(OH)$_2$ (150 g/liter)
0.1 $\underline{N}$ NaOH
H$_3\overline{PO}_4$ – MnSO$_4$ mixture‡
0.02 $\underline{M}$ KMnO$_4$

Period 5

Apparatus for filtration plating (Figure 18)
Counter for radioactivity
1 $\underline{M}$ BaCl$_2$

PROCEDURE

LABORATORY SCHEDULE

Period 1

Each pair of students will receive two rats. Feed, water, and weigh the rats daily. Continue the feeding and weighing for several days. This is done in order to become familiar with the handling of rats and to train the rats not to be excited by the commotion in the laboratory or by the handling. The rats are tame and will not bite if picked up without fear and handled gently. Pet the rat and make it feel at ease. It is important to have the rat calm at the time of the experiment; a frightened rat does not form glycogen. Twenty-four hr prior to period 2, remove all food from the cage to initiate the fast.

---

*Preparation of Lactobacillus casei. Medium for growth of the bacteria: 0.5% Difco yeast extract; 0.5% Difco tryptone; 1.0% glucose; 0.6% sodium acetate·3H$_2$O; and 0.5% K$_2$HPO$_4$. Adjust the medium to pH 6.3 – 6.5 and autoclave it in 2 liter lots in 3 liter Erlenmeyer flasks stoppered with cotton plugs (for preparation of about 3 g of bacteria/2 liters). Also sterilize for each 2 liter lot of medium one 125 ml Erlenmeyer flask that contains 50 ml of medium. These small flasks will be used for inoculation of the large flasks. Inoculate each 125 ml flask from the L. casei culture, and incubate 24 hr at 37°. Then inoculate the 2 liter lots of medium with the contents of the 125 ml flasks and incubate the large lots at 37° for 36 hr. Harvest the cells in a Sharples centrifuge or an International centrifuge. Wash the cells with saline and then distilled water and finally suspend them in an equal volume of distilled water.

† Add the 2,4-dinitrophenylhydrazine to hot 2$\underline{N}$ sulfuric acid and stir and heat until it dissolves. A precipitate forms if the solution is stored; this should be filtered off before the solution is used.

‡ Dissolve 100 g of MnSO$_4$·4H$_2$O and 25 ml of 85% H$_3$PO$_4$ in water and dilute the solution to 1 liter.

Period 2

A. Administration of glucose solutions by stomach tube and the collection of respiratory CO_2 for 3 hr.

B. Anesthesia; removal of the liver; isolation and hydrolysis of the glycogen.

Period 3

C. Determination of reducing sugars in the glycogen hydrolysate.

D. Bacterial fermentation of the glucose to lactic acid.

Period 4

E. Oxidation of the lactic acid to acetaldehyde and CO_2.

Period 5

F. Determination of radioactivity of the samples.

A. *Administration of Glucose Solutions and Collection of Respiratory Carbon Dioxide*

Preparation of the respiration apparatus. Set up the apparatus illustrated in Figure 19 the day before it is to be used, and check it for leaks with water in the collection tubes. Two sets of the apparatus will be needed for each pair of students.

At the beginning of period 2, measure approximately 30 ml of 1 N NaOH into each test tube. Quickly stopper the tubes tightly. Turn the vacuum on slowly until air begins to bubble through the system. Check to be sure that the tubes have been hooked up properly. Increase the speed of aeration slowly until the bubbles form a little faster than you can count them. The bubbles should come out of the last tube at the same rate as they are formed in the first tube. If they do not, there is a loose connection which causes a leak. When the system is tight, turn off the vacuum and then remove the stopper from the mason jar. You are now ready to reweigh the rats and administer the glucose and fatty acids.

Feeding the rat by stomach tube. The instructor will demonstrate the technique of feeding a rat by stomach tube; thereafter, it is to be done by the students. It is highly advisable to practice feeding unlabeled glucose solutions for several days before attempting to feed rats in the experiment. Rats for such practice will be available.

The C^{14} compounds have a relatively high radioactivity (2 μc per millimole) and they must be handled carefully. Be sure to follow instructions given on page 141 concerning the handling of C^{14} compounds. Once the C^{14} has been fed, the radioactive carbon becomes diluted by normal carbon from the animal body and subsequent precautions are not so critical. One of the rats will receive 0.5 g of glucose and 2.5 millimoles of non-labeled fatty acid per 100 g body weight; the other rat will receive 0.5 g of glucose and 2.5 millimoles of labeled fatty acid that contains 5 μc per 100 g body weight. The stomach feeding of glucose and fatty acids is done

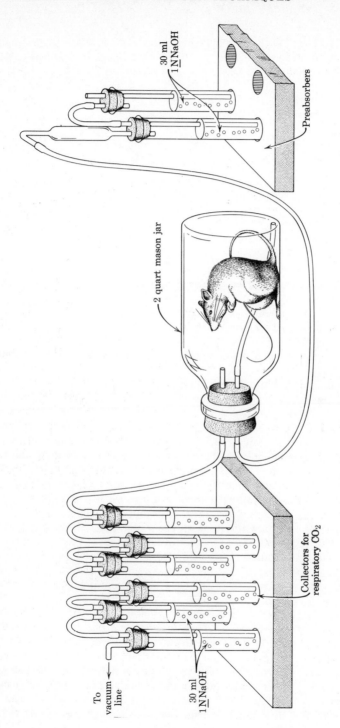

Figure 19. Apparatus for metabolic studies in the rat.

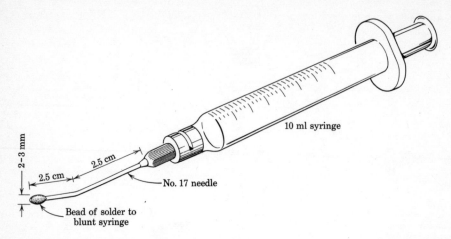

Figure 20. Syringe and stomach tube for administration of radioactive compounds to rats.

with a 10 ml syringe with a blunt needle (Figure 20). The rat is grasped by the tail and is allowed to try to pull away on the table. While it is doing this, grasp it well down the sides of the neck with the thumb and second finger of the left hand. The index finger is placed at the back of the head. If the skin is pulled up taut, the rat cannot turn its head. One student will hold the hind legs of the rat firmly so that the rat's body is straight. The syringe is held almost parallel with the body and with the top of the tube against the roof of the mouth. The tip is gently pushed along the roof of the mouth and down the esophagus. About two-thirds of the way down an obstruction may be encountered. If the syringe is gently rotated, the tube will slip by this point. The glucose solution is slowly injected while giving the rat a chance to swallow, and the tube then is gently removed. Operate as rapidly as possible for the rat cannot breathe while the tube is in its esophagus. The rat receiving non-labeled compounds should be fed first so the student will gain familiarity with the technique; for the same reason, all succeeding operations also should be done first on the sample that contains the non-labeled compound. Results obtained with the non-labeled compound will serve as a control. As soon as the solution has been injected, the rat is placed in the CO_2 collection apparatus. Record the amounts fed in Tables XV and XVI (pages 165–166).

Collection of respiratory CO_2. Place a sheet of folded paper toweling in the mason jar to absorb urine. Then face the rat into the jar with the jar opening tilted down. The rat will usually crawl into the jar if it is so tilted. Stopper the jar tightly while it is tilted. If you tilt the jar opening up or level the jar, the rat often will try to crawl out. Cover the jar with a towel, for the rat will remain quieter in the dark. Start air gently bubbling through the alkali train until bubbles form somewhat faster than you can count them. The flow must be rapid or the CO_2 concentration will increase in the bottle. This in turn will stimulate the respiratory centers and cause rapid breathing and excitation. Check to see that the system is airtight and that the flow rate is the same in the preabsorbers as in the collection tubes. If for some reason the air flow is discontinued and cannot be started, remove the stopper and place the animal under the hood so that the animal will not suffocate and the C^{14} from the respired air will not go into the laboratory. Aeration is maintained for 3 hr; meanwhile, prepare for the removal and digestion of the liver as described in the next section of this experiment.

B. *Anesthesia: Removal of the Liver and Isolation and Hydrolysis of the Glycogen*

Before the end of the 3 hr collection period, fill a 400 ml beaker two-thirds full of hot water, and bring the water to a boil. Place a 40 ml centrifuge tube that contains 10 ml of 30% KOH solution into this bath 5 min before the collection period is to end. Place a marble over the opening of the tube to prevent loss by evaporation. Cover the desk top on which the operation is to be done with paper. Lay out paper toweling, scissors, forceps, and dissecting board.

Anesthesia with ether. At the end of the 3 hr collection period, shut off the vacuum, and remove the rubber tubing from the glass tubes in the mason jar. Immediately soak a pad of cotton with ether. Remove the stopper from the mason jar, drop in the cotton, and replace the stopper. CAUTION: be sure that work with ether is well removed from all open flames (e.g., flames for the water baths). The rat will be anesthetized in about 5 min. In the meantime, attach the rubber tubing from the inlet of the first gas scrubber tube to the outlet of the last tube, thus making a closed system and protecting the alkali from the CO_2 of the air. Likewise, if you protect the preabsorbers from air, they may be used without change in the latter part of the experiment.

Watch the rat carefully and, as soon as it does not respond when the bottle is shaken, remove the rat from the jar. Test the rat by pinching the foot sharply or test the eye reflex by touching the eyeball with your forceps to determine whether or not the animal is conscious. If the rat responds, place it in the jar again for a short time.

Removal of the liver. Use rubber surgical gloves. Place the rat on the dissecting board covered with paper towels and cover the head with the ether-soaked cotton. Then open up the belly, exposing the peritoneal cavity. To do this, place the rat on its back, and, with the finger or forceps, pull up the skin on the lower part of the abdomen. Cut through the skin and muscle layer until you have exposed the intestines. Then, with the skin raised, cut forward down the mid line; avoid cutting the internal organs and stop anteriorly at the diaphragm. Do not cut into the thoracic cavity, or breathing will stop. Now cut down each side just below the ribs so the opening is enlarged. The liver may be picked up with the forceps by one partner while the other snips off the attached membranes. Remove the liver as quickly as possible, for glycogen is destroyed rapidly in the surviving liver tissue. Do not try to do a delicate job of surgery; pull and snip it out. Blot the blood off the liver on paper toweling, cut the liver into $\frac{1}{2}$ in.[2] pieces, and <u>immediately</u> place them in the tube of alkali. Put the tube in the water bath, and replace the marble.

Wrap the rat carcass in paper, and place it in the receiver marked for this purpose. Wash the gloves thoroughly to remove any C^{14}. Remove the gloves and wash your hands thoroughly with soap and water to remove any C^{14} that may be on them. Also, clean up any blood that may have dropped by accident on the desk or floor and dispose of it in the receiver for the rats.

Isolation of the glycogen. The liver is to be digested until the solution is an opaque burgundy red and large particles of tissue are absent. This digestion will require $\frac{1}{2}$ to 1 hr. For the first 5 min stir the mixture well with a stirring rod; then repeat the stirring at 10 min intervals. Occasionally examine the solution against the light. When it is free of particles, the digestion is complete.

Remove the tube, cool the contents to $40-50°$, and add 18 ml of 95% ethanol. A white flocculent precipitate will appear, which is glycogen with some impurities of a protein nature. Stir the mixture thoroughly, and place the tube in a water bath at about $80°$ for 5 min. (<u>CAUTION</u>: extinguish the burner flame while the alcoholic solution is in the bath.) Remove the tube and cool it to room temperature.

Centrifuge the tube for 10 min at 2000 rpm and pour off the supernatant fluid into the receptacle that is provided for liquid radioactive wastes. The glycogen with protein impurities will be the yellowish-brown residue at the bottom of the tube. Add 15 ml of 10% trichloracetic acid (TCA) in this way: first, add about ten drops and stir until the solid material is dispersed into a smooth paste; then, add the remaining TCA in 3 portions. Stir the mixture vigorously after each addition. The glycogen will go into solution, and the protein will precipitate. The solution should have an opalescent appearance because of the presence of glycogen.

The mixture is centrifuged for 10 min, and the supernatant fluid is then carefully poured down a stirring rod into a second centrifuge tube (be sure all tubes are clearly labeled). <u>Save</u> the <u>supernatant</u> <u>solution</u>, and discard the precipitate into the receptacle for solid wastes. To the supernatant solution add 25 ml of 95% alcohol and mix thoroughly with a stirring rod. A white flocculent precipitate, which is fairly pure glycogen, should come down. Centrifuge for 10 min at 2000 rpm and discard the supernatant fluid into the receptacle for liquid wastes.

Another precipitation is done to remove any TCA which may adhere to the glycogen. To do this, dissolve the glycogen in 5 ml of distilled water that has been brought to boiling. Place the tube in the boiling water bath, and stir to disperse solid material and to dissolve the glycogen. Continue to stir until the solid material is either dissolved or dispersed in solution, leaving no solid gelatinous material at the bottom of the tube. Add 5 ml more of hot distilled water; then add 12 ml of 95% alcohol while mixing with the stirring rod. A white flocculent precipitate of glycogen that is quite pure will come down. Centrifuge for 10 min at 2000 rpm and discard the supernatant fluid. Invert the tube and allow it to drain; then place it in the hot water bath for 2 min to remove all traces of alcohol.

Hydrolysis of the glycogen. The glycogen should be hydrolyzed before the next period by heating it in a boiling water bath with 2.0 ml of 1.0 N H_2SO_4 for $1\frac{1}{2}$ hr. The tube is to be shaken occasionally during the hydrolysis. Also cap the tube with a marble during the hydrolysis and check occasionally to be sure that the water has not evaporated. Loss of water will lead to charring of the sample by the remaining acid.

C. *Determination of the Reducing Sugars in the Hydrolyzate*

While the amount of reducing sugar is being determined, one student may prepare for the bacterial fermentation to be performed in part D. The fermentation should not be started, however, until the glucose determination is completed and it is decided whether or not carrier glucose should be added. A carrier is added when the amount of material available is too small to permit accurate study. Since the isotopic sugar is chemically the same as the carrier sugar, the isotopic sugar is "carried" along in the reactions with normal sugar. In this way, minute amounts of radioactive materials may be handled.

To the hydrolyzate add a spatula tip (matchhead) amount of decolorizing carbon, stir well, and let the mixture stand for 10 min. Then add 4 ml of distilled water and filter the mixture into a volumetric flask graduated at 10 ml. Use a small filter paper (7 cm) and wet the funnel so that the paper adheres to the sides. Guide the solution onto the filter with a stirring rod. Rinse the tube and stirring rod with two 1.5 ml portions of water. Add water to make the filtrate volume exactly 10 ml.

Pipet 0.10 ml of the hydrolyzate into a 10-ml volumetric flask and dilute it to 10 ml with water. Determine the reducing sugar quantitatively with 1 ml aliquots of this dilution. Set up a blank and standards as outlined in Table XIV. From your data, calculate the amount of glucose present in the hydrolyzate. If you have less than 0.15 millimoles of glucose, consult your instructor. It may be necessary to add glucose as a carrier in order to have sufficient material for the fermentation.

D. *Bacterial Fermentation of Glucose to Lactic Acid* (5, 6)

The fermentation is carried out with a 36 hr washed suspension of Lactobacillus casei under anaerobic conditions in bicarbonate buffer. The bacterial fermentation consists essentially of the reaction: glucose → 2 lactic acid. The rate and extent of fermentation will be followed by the decrease in glucose content as described below. (Alternatively, the fermentation may be followed by the increase in gas volume by CO_2 released in the reaction: $NaHCO_3 + CH_3 \cdot CHOH \cdot COOH \longrightarrow CH_3 \cdot CHOH \cdot COONa + CO_2 + H_2O$.)

The fermentation will be carried out in 125 ml Erlenmeyer flasks suspended in a water bath at 35°. Pour the remaining 9.9 ml of hydrolyzed glycogen solution into the 125 ml flask; rinse the volumetric flask with 5 ml of water and add this to the flask. Now add 5 ml of 0.8 M $NaHCO_3$. This will neutralize the sulfuric acid used in the hydrolysis and provide an excess of bicarbonate to react with lactic acid formed in the fermenta-

TABLE XIV. PROTOCOL FOR THE DETERMINATION
OF REDUCING SUGARS

Tube number	Blank	S_1	S_2	S_3	U_1	U_2
H_2O	7	6	5	3	6	6 ml
Hydrolyzed glycogen; 0.1 ml diluted to 10 ml	0	0	0	0	1	1 ml
Standard, 0.12 mg/ml	0	1	2	4	0	0 ml
Reagent A + B*	2	2	2	2	2	2 ml

Mix the tube contents and heat exactly 10 min in a boiling-water bath. After the 10 min heating period, immediately cool the tubes in a beaker of cold water. Add the arsenomolybdate reagent after the tubes are at room temperature.

Arsenomolybdate reagent	1	1	1	1	1	1 ml

Mix thoroughly until all the precipitate is dissolved, then dilute to 10 ml with water and mix the contents. Read the color intensity in a colorimeter at 540 mμ within $10-30$ min.

Colorimeter reading						
Mg of glucose	0	0.12	0.24	0.48		

* Mix 25 ml of reagent A with 1 ml of reagent B on the day that the reagent is to be used.

tion. Add 1 drop of 0.2% bromthymol blue; the pH should be such that the color of this indicator is blue-green. Add 2 ml of distilled water, or, if the glucose yield is low, a 2 ml solution of carrier glucose. Finally add 3 ml of a 50% suspension of Lactobacillus casei.

Loosely stopper the apparatus with a cork or cotton plug and place the flask on the shaker in the water bath. Insert through the stopper a piece of 3 mm glass tubing (loose fit) which reaches to the bottom of the flask. CO_2 will be introduced into the vessel through this tubing and it should bubble through the solution somewhat faster than you can count the bubbles. Gas the solution for at least 5 min and shake the flask continuously to facilitate saturation of the solution with CO_2.

Discontinue the gas flow but continue to shake the flask, and follow the course of the fermentation in a rough way by withdrawal of appropriate aliquots for measurement of residual reducing sugar by the method described in part C of this experiment. When the reducing sugar content has dropped to a negligible or to a small but constant amount, the flask may be removed for further treatment of the lactic acid.

E. *Oxidation of the Lactic Acid to Acetaldehyde and Carbon Dioxide*

The lactic acid will be oxidized with $KMnO_4$ to acetaldehyde and CO_2:

$$CH_3 \cdot CHOH \cdot COOH \xrightarrow{KMnO_4} CH_3 \cdot CHO + CO_2 + H_2O$$

This reaction is the basis of a well-established method for the quantitative determination of lactic acid (7). The CO_2 will be absorbed in alkali scrubbers and the aldehyde will be trapped as the insoluble 2,4-dinitrophenyl-hydrazone, an orange-yellow compound, which is easily crystallized from water-alcohol mixtures and melts at 164°.

Set up the 29 by 200 mm test tubes and flask as illustrated in Figure 21. The preabsorber tubes which you previously used may be used again without change of alkali, if they have been protected from CO_2. Filter the 0.5% 2,4-dinitrophenylhydrazine if it is not clear, and add 30 ml of the reagent to each of the first two collector tubes. Add 30 ml of distilled water to the other two tubes and mark them with your wax pencil at the 30 ml level. Empty the water and fill the tubes to the marked level from the stock $Ba(OH)_2$ solution. Stopper the tubes immediately, and, with the 100 ml round-bottom flask in place, test for leaks by passing air through the train. The rate of bubbling should be the same in the preabsorbers as in the collector tubes. Remove the flask and connect the first collector tube with the last collector tube to protect the $Ba(OH)_2$ solution from CO_2 in the air.

When the bacterial fermentation is complete, pour the mixture into a 40 ml centrifuge tube (labeled), wash out the flask with 5 ml of water, and centrifuge the combined mixture and wash for 10 min at 2500 rpm. Decant the supernatant solution into the 100 ml round-bottom flask. If the fermentation for some reason was weak and gave a poor yield of lactate, it may be necessary to add carrier lactate at this point.

To the clear solution add 10 ml of the H_3PO_4–$MnSO_4$ mixture and bring the solution to a boil for 5 min. The solution should be acid (bluish-purple) to Congo red so that boiling will drive off all the CO_2 liberated from the bicarbonate. It is necessary to free the solution of CO_2; otherwise this CO_2 will contaminate the radioactive CO_2 which will be formed during the oxidation.

Cool the solution in tap water. When it is cool, place the flask in

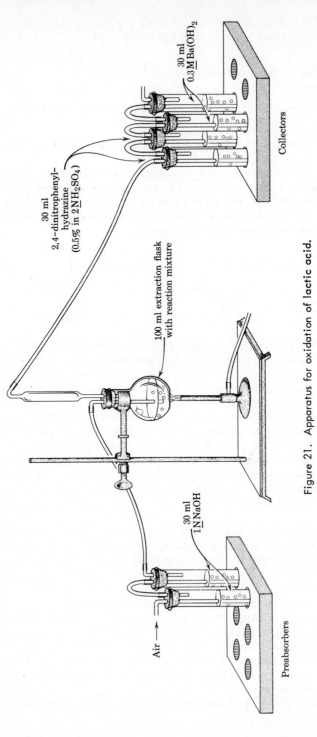

Figure 21. Apparatus for oxidation of lactic acid.

your oxidation train. When everything is in order so that aeration may be started without delay, add 25 ml of 0.1 M KMnO$_4$ from a graduate cylinder. Quickly insert the stopper firmly into the flask. Now slowly turn on the air and adjust the rate to about 3−4 bubbles per second.

The mixture should be heated with a low flame and brought to a boil in about 5 min. Boil only until condensate forms a drop in the outlet tube of the reaction flask, then discontinue heating for 10 min. After that time, again bring the solution to a boil for a few seconds. Aerate for 10 min more, and the reaction is complete.

If all goes well, the dinitrophenylhydrazone and BaCO$_3$ will begin to form shortly after the mixture is brought to a boil. Practically all of the hydrazone should be formed in the first tube and the BaCO$_3$ in the third. At the conclusion of the reaction remove the stopper from the reaction flask and then turn off the air. Hook up the first and last collector tubes to protect the train from air. These tubes are to be stored until you are ready to plate the contents.

F. *Determination of Radioactivity of the Samples*

Radioactivity determinations will be done on samples from both the C^{14}-fed animal and the non-radioactive, control animal. The alkali scrubbers that contain the respiratory CO$_2$ will be combined and diluted to 500 ml. A 1 ml aliquot will be taken for precipitation of the carbonate as BaCO$_3$ on a weighed filter pad by the method of filtration plating (Experiment 35). From the weight and activity of the BaCO$_3$ the amount and activity of the respiratory CO$_2$ can be calculated.

The 2,4-dinitrophenylhydrazone* precipitates should be well washed with water and may be plated by filtration from water suspension onto weighed filter pads. Alternatively, the washed precipitates may be dissolved in acetone and plated by evaporation on weighed sample pans (see Experiment 34 for details of the general procedure for plating by evaporation). The sample should be dried gently under a light bulb before it is reweighed and counted.

The BaCO$_3$ samples from the lactate degradation are centrifuged. The supernatant fluid is decanted, and the BaCO$_3$ is washed with water and centrifuged again. Then it is suspended in a little water and plated by filtration as above. The weighed plate is counted. Apply appropriate corrections for background, coincidence, and self-absorption to all samples.

TREATMENT OF DATA

Assume that you wish to publish the results of your experiment in the Journal of Biological Chemistry. Draw up a manuscript in the acceptable form and style as set forth on the first page of any recent number of that journal. Show clearly how your results relate to currently accepted

---

* The dinitrophenylhydrazones of these aldehydes are insoluble in water, slightly soluble in cold alcohol, fairly soluble in hot alcohol, and very soluble in acetone.

patterns for the metabolism of fats and carbohydrates in the mammal.

Pay particular care to the manner in which radioactivity is reported. Values for the precision of counting and a statement of the significance of observed radioactivity values should be included.

QUESTIONS

1. There are many chemical schemes for complete and specific degradation of glucose other than the one used in this experiment. Devise or find in the chemical literature two alternative schemes.

2. Why was glycogen released from the liver tissue by alkaline degradation rather than acid degradation?

3. It has been assumed in this experiment that Lactobacillus casei metabolizes glucose solely to lactic acid. From a brief search of the literature, you should be able to ascertain the reliability of this assumption.

4. Why was glucose administered to the animal at the same time that the labeled fatty acid was given?

References

1. Lifson, N., Lorber, V., Sakami, W., and Wood, H. G. 1948. Incorporation of Acetate and Butyrate Carbon into Rat Liver Glycogen by Pathways Other than Carbon Dioxide Fixation. J. Biol. Chem., 176, 1263–84.

2. Lipmann, F. 1948. Biosynthetic Mechanisms. In The Harvey Lectures (1948–49), 99–123.

3. Krebs, H. A. 1948. The Tricarboxylic Acid Cycle. In The Harvey Lectures (1948–49), 165–99.

4. Ochoa, S. 1950. Enzyme Studies in Biological Oxidations and Synthesis. In The Harvey Lectures (1950–51), 153–80.

5. Wood, H. G., Lifson, N., and Lorber, V. 1945. The Position of Fixed Carbon in Glucose from Rat Liver Glycogen. J. Biol. Chem., 159, 475–89.

6. Abraham, S., Chaikoff, I. L., and Hassid, W. Z. 1952. Conversion of C^{14} Palmitic Acid to Glucose. J. Biol. Chem., 195, 567–81.

7. Friedemann, T. E., and Graeser, J. B. 1933. The Determination of Lactic Acid. J. Biol. Chem., 100, 291–308.

TABLE XV. DATA SHEET FOR EXPERIMENT 36

Data Required	Rat Fed C^{14} in the Form of	Control Rat

Amounts fed and recovered:
  Weight of rat
  Glucose fed (mM)*
  Fatty acid fed (mM)
  Sugar from glycogen (mM)
  Glucose added as carrier (mM)
  Lactate added as carrier (mM)		

Respiratory CO_2:
  Weight of $BaCO_3$ (mg)
  Cts/min minus background
  Correction factor
  Corrected cts/min
  Cts/min/mM		

Carbons 3, 4:
  Weight of $BaCO_3$
  Cts/min minus background
  Correction factor
  Corrected cts/min
  Cts/min/mM		

Carbons 1, 2, 5, 6:
  Weight of dinitrophenylhydrazone
  Cts/min minus background
  Correction factor
  Corrected cts/min
  Cts/min/mM		

Original fatty acid:
  Cts/min minus background
  Correction factor
  Corrected cts/min
  Cts/min/mM		

* In this Table, the abbreviation mM stands for "millimole(s)."

TABLE XVI. SUMMARY OF RESULTS OF EXPERIMENT 36

Data Required	Rat Fed C^{14} in the Form of	Control Rat
Compound fed: mM* fed Cts/min/mM Total counts fed		
Respiratory CO_2: mM collected Cts/min/mM Total counts collected % of fed radioactivity % of respired CO_2 from fatty acid		
mM of glycogen as glucose		
Carbons of 3, 4 positions: Cts/min/mM % of fed radioactivity		
Carbons of 1, 2, 5, 6 positions: Cts/min/mM % of fed radioactivity		

*In this Table, the abbreviation mM stands for "millimole(s)."

EXPERIMENT 37. CARBON DIOXIDE FIXATION BY PROPIONIC ACID BACTERIA* (6 periods; usually done by pairs of students)

OBJECTIVE

The experiment permits the student to obtain a balance of radioactive reactants and products of bacterial metabolism.

PRINCIPAL EQUIPMENT AND SUPPLIES

For Preparation of Inoculum

Autoclave
pH meter
Culture of Propionibacterium pentosaceum (ATCC 4875), or Propioni-
bacterium rubrum (ATCC 4871), or Propionibacterium jensenii (ATCC
4867)
Sodium lactate (85% syrup)
Difco yeast extract
$MgSO_4 \cdot 7H_2O$
$CaSO_4 \cdot 2H_2O$
Agar
Phenol red
Cysteine·HCl

For Fermentation

30° incubator
Glycerol
0.1 N NaOH
Sodium carbonate solution (0.01−0.001 M) containing 2 μc C^{14}/ml
1 N H_2SO_4

For Determination of CO₂ Fixed

Counter suitable for C^{14}
Diffusion vessel (1)
1 N NaOH (CO₂-free)

For Fractionation of the Fermentation Mixture

All-glass steam distillation apparatus
Metal bath
Thermocouple and pyrometer for measurement at 450°
Centrifuge
Pyrolysis tube and diffusion apparatus [p. 93 of ref. (2)]

---

* This experiment was designed by Dr. H. A. Barker of the University of California, Berkeley, California.

Soxhlet extractor
Vacuum pump
Dry ice — acetone bath
Safety glasses
Oxalate-washed Whatman No. 1 paper
Bromphenol blue spray (3)
1 N lactic acid
$0.\overline{25}$ N $Ba(OH)_2$
0.1 $\overline{N}$ HCl

PROCEDURE

The propionic acid bacteria were the first typically heterotrophic organisms shown to utilize carbon dioxide. Werkman and Wood and others demonstrated that in the fermentation of glycerol, lactate, and other substrates by these bacteria, carbon dioxide is incorporated into the carboxyl groups of propionic and succinic acids (4, 5, 6).

In this experiment, Propionibacterium pentosaceum or other propionic acid bacterium is allowed to ferment glycerol in the presence of C^{14}-labeled bicarbonate. About 70% of the glycerol is converted to propionic acid and water (7), while small amounts of acetate and succinate also are formed. You will determine the initial and final amounts and initial and final specific activities of the carbon dioxide. The propionic and succinic acids will be separated from the fermented medium, and the propionic acid will be decarboxylated to demonstrate the presence of C^{14} in the carboxyl group.

An outline of the procedure is contained in Diagram 2. Certain details of the steps will be elaborated below, but reference to the original literature should be made for complete details in every instance.

1. Inoculum

The inoculum should be grown in a cotton-plugged test tube half-filled with the following medium: 0.5 vol.% sodium lactate (85% syrup), 0.5% Difco yeast extract, 3 vol.% of 1 M potassium phosphate buffer of pH 7.4, 0.01% $MgSO_4 \cdot 7H_2O$, 0.005% $Ca\overline{SO_4} \cdot 2H_2O$, 0.2% agar, 0.5 vol.% of a 0.04% aqueous solution of phenol red indicator, and glass-distilled water. Boil the medium briefly to dissolve the agar. Just before sterilizing the medium, add 0.03% cysteine·HCl dissolved in about 1 ml of water and neutralize the medium with NaOH to pH 7. Autoclave for 20 min at 15 lbs/in.^{2} (psi). Cool and then inoculate the tube and incubate for 48 hr at 28–32°. The culture is then ready to use as an inoculum for the main experiment.

2. Fermentation of Glycerol and Fixation of $C^{14}O_2$

In a clean 200 ml Erlenmeyer flask, place 1 millimole (106 mg) of sodium carbonate dissolved in about 4 ml of distilled water. Add 0.5 ml of 0.1 N NaOH and 1–2 ml of a dilute (0.01–0. 001 M) sodium carbonate solution containing a total of 1–2 μc of C^{14}. The flask containing the

DIAGRAM 2. SCHEME FOR INCORPORATION OF C^{14} INTO
PROPIONIBACTERIUM PENTOSACEUM AND FOR
FRACTIONATION OF THE RADIOACTIVE PRODUCTS

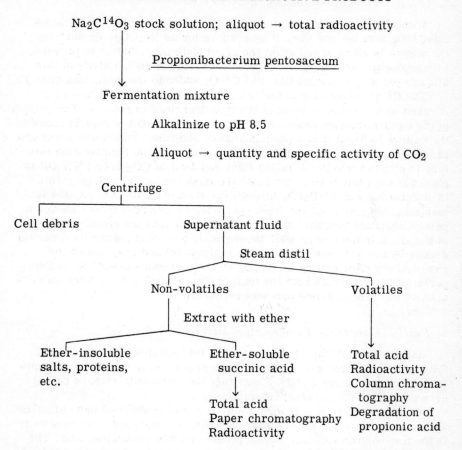

$Na_2C^{14}O_3$ stock solution; aliquot → total radioactivity

Propionibacterium pentosaceum

Fermentation mixture

Alkalinize to pH 8.5

Aliquot → quantity and specific activity of CO_2

Centrifuge

Cell debris   Supernatant fluid

Steam distil

Non-volatiles   Volatiles

Extract with ether

Ether-insoluble   Ether-soluble   Total acid
salts, proteins,   succinic acid   Radioactivity
etc.   Column chroma-
                         Total acid   tography
                         Paper chromatography   Degradation of
                         Radioactivity   propionic acid

carbonate is closed with a cotton plug and is then autoclaved for 5 min at
15 psi. At the same time, 150 ml of a medium similar to that described
under "inoculum" above, except that the lactate is replaced by 0.3 vol.%
glycerol and the agar is omitted, is autoclaved in a 250-ml flask. A test
tube containing 3 ml of $1 N H_2SO_4$ is also autoclaved. The next operations
are done under a hood. After the solutions have been sterilized and cooled,
the glycerol-yeast extract medium is added, using aseptic technique, to
the flask containing the labeled carbonate. With a sterile 1-ml pipet, adjust the pH of the medium to pH 7.4 – 7.6 by addition of sterile H_2SO_4;
approximately 1 ml is required. Then inoculate the medium with 2 ml of
the sediment from the semisolid lactate culture with another sterile pipet.
Swirl the flask to mix the contents thoroughly. Finally, push the sterile
cotton plug down into the neck of the flask, and seal the flask tightly with

a rubber stopper. Incubate the culture for 3–4 days at 30°, shaking it occasionally. Judge the cessation of growth by observing turbidity.

3. *Determination of the Amount of CO_2 Fixed*

Withdraw a small aliquot (calculate the proper amount) of the stock $Na_2C^{14}O_3$ solution and plate it by evaporation as Na_2CO_3. Be sure that the aliquot is alkaline and count the dried plate immediately to prevent exchange with carbon dioxide of the air. From the radioactivity of this aliquot, you may calculate the total $C^{14}O_2$ added to the incubation mixture.

The $C^{14}O_2$ remaining at the end of the incubation period may be recovered with a diffusion flask of the type described in ref. (1). The stopper of the culture flask is removed and CO_2-free 1 N NaOH is rapidly added to the culture to bring it to a phenolphthalein endpoint. This alkali serves to prevent loss of carbon dioxide. A 10-ml aliquot of the fermentation mixture is pipetted into the diffusion flask and 1 ml of CO_2-free 1 N NaOH is placed in the center-well. Evacuate the flask through a syringe needle; then inject 2 ml of 1 N H_2SO_4 through the syringe needle and into the fermentation aliquot. Rock the flask gently for 30 min on a rotary shaker to permit complete diffusion of the carbon dioxide into the alkali. Aliquots of the alkali in the center-well subsequently are analyzed for total carbon dioxide by the Van Slyke gasometric analysis (8) and are counted for radioactivity after plating as $BaCO_3$ by the procedure described in Experiment 35. Calculate both the total carbon dioxide and the total radioactivity as carbon dioxide that was recovered.

4. *Fractionation of the Fermentation Mixture*

Adjust the pH of the remainder of the fermentation mixture to pH 2 (pink to thymol blue) with dilute H_2SO_4 and aerate the mixture for 10 min in the hood to remove $C^{14}O_2$. Centrifuge the mixture to remove cell debris and decant the supernatant solution.

An aliquot of the supernatant solution is steam distilled in an all-glass apparatus to separate volatile and non-volatile components. Radioactivity in the non-volatile fraction will be present mainly in succinic acid. The non-volatile residue in the distillation pot may be dried and extracted with ether in a Soxhlet extractor. The radioactivity and acid content of the ether extract may be determined, and the succinic acid may be identified by paper chromatography (3).

Radioactivity in the volatile fraction represents the main fraction of the fixed $C^{14}O_2$, and it will be present mainly in propionic acid. Titrate an aliquot of the volatile fraction for total acid content. Another aliquot of the volatile fraction may be analyzed by partition chromatography of the liquid-liquid (Experiment 6) or gas-liquid (Experiment 7) types; you may need to add carrier acetic acid for this analysis if you wish to test the acetic acid fraction for radioactivity.

5. *Degradation of Propionic Acid*

The location of the C^{14} in the propionic acid of the volatile fraction may be determined by decarboxylation of the barium salt according to the reaction

$$(CH_3CH_2COO)_2 \ Ba \xrightarrow{450°} (CH_3CH_2)_2CO + BaCO_3$$

The carbon in the barium carbonate is known to come mainly from the carboxyl group of the fatty acid (9). Draw an aliquot of the volatile fraction that contains about 200 micromoles of propionic acid and neutralize it to pH 7 with $0.25 \underline{N} Ba(OH)_2$. Concentrate the mixture to dryness on the steam bath in an evaporating dish. Dissolve the residue in the minimum volume of water, centrifuge to remove any $BaCO_3$, and transfer it to one tube of the diffusion apparatus described on page 93 of ref. (2). Finally, dry the tube and contents in an oven at 105°. When the salt is thoroughly dry and cool, apply silicone grease to the joint and attach the tube to the diffusion apparatus. Evacuate the apparatus with a good vacuum pump (preferably an oil pump), close the stopcock, and disconnect the pump. Now insert the end of the arm containing the barium propionate into a metal bath heated to 450°, while the opposite arm is cooled in a dry ice— acetone bath. Within 1 min the salt begins to decompose and within 2–3 min the reaction is complete. (<u>CAUTION</u>: It is advisable to wear safety glasses during this operation.)

6. *Separation of Carbon Dioxide from Propionate*

When the reaction tube is cool, disconnect it from the diffusion apparatus and insert a small test tube of about 4 ml capacity in the reaction tube. Pipet 2 ml of $1 \underline{N}$ lactic acid into the insert tube and reconnect the larger tube to the diffusion apparatus. Remove and clean the receiver arm and place in it 3 ml of $0.25 \underline{N} Ba(OH)_2$ containing 1 drop of phenolphthalein indicator. Evacuate the apparatus on a water pump until the solutions begin to boil. Be careful to avoid letting the lactic acid in the insert tube come in contact with the barium carbonate in the large tube. Now close the stopcock, disconnect the apparatus from the pump, and carefully tip it so that the acid comes in contact with the barium carbonate. Shake the apparatus by hand continuously for 5–10 min until the CO_2 has been absorbed in the $Ba(OH)_2$. Now release the vacuum, remove the tube containing the barium hydroxide, and titrate the excess alkali to pH 8.3 (faint pink with phenolphthalein) with $0.1 \underline{N} HCl$. Transfer the resulting suspension of $BaCO_3$ in barium chloride solution to a 5 ml test tube and centrifuge it. Wash the precipitate twice with water and once with 95% ethanol. Transfer a suitable sample of the alcoholic slurry of the barium carbonate to a counting disc and determine the specific activity of the compound. Calculate the percentage of the total C^{14} that is in the carboxyl group of the propionic acid.

The accumulated data at this point should permit you to calculate a complete balance of $C^{14}O_2$ fixation and C^{14} distribution in the various fractions of the fermentation mixture. Also, you can compare the total carbon dioxide fixed during the fermentation with the total titratable acidity of both the volatile and the ether-extractable, non-volatile fractions.

TREATMENT OF DATA

Summarize your results and deliver a formal, oral report to the class.

QUESTIONS

1. What are heterotrophic and autotrophic organisms?
2. What are the paths of CO_2 fixation in the Propionibacteria that would lead to the radioactive products you observed?
3. Would you expect appreciable radioactivity to remain in the cell debris upon centrifugation of the fermentation mixture? How could you count small levels of radioactivity in such a sample?

References

1. Katz, J., Abraham, S., and Baker, N. 1954. Analytical Procedures Using a Combined Combustion-Diffusion Vessel. Anal. Chem., 26, 1503–4.
2. Calvin, M., Heidelberger, C., Reid, J. C., Tolbert, B. M., and Yankwich, P. F. 1949. Isotopic Carbon. John Wiley and Sons. New York.
3. Block, R. J., Durrum, E. L., and Zweig, G. 1955. Paper Chromatography and Paper Electrophoresis, A Manual. Academic Press. New York.
4. Werkman, C. H. 1951. Assimilation of Carbon Dioxide by Heterotrophic Bacteria. In Bacterial Physiology, C. H. Werkman and P. W. Wilson, editors. Academic Press. New York.
5. Werkman, C. H., and Wood, H. G. 1942. Heterotrophic Assimilation of Carbon Dioxide. In Advances in Enzymology, 2, 135–82. Interscience Publishers. New York.
6. Carson, S. F., and Ruben, S. 1940. CO_2 Assimilation by Propionic Acid Bacteria Studied by the Use of Radioactive Carbon. Proc. Natl. Acad. Sci. U.S., 26, 422–6.
7. Leaver, F. W., Wood, H. G., and Stjernholm, R. 1955. The Fermentation of Three Carbon Substrates by Clostridium Propionicum and Propionibacterium. J. Bact., 70, 521–30.
8. Van Slyke, D. L., and Neill, J. M. 1924. The Determination of Gases in Blood and Other Solutions by Vacuum Extraction and Manometric Measurement. J. Biol. Chem., 61, 523–73. See also, ibid., 73, 121–6.
9. Wood, H. G., Werkman, C. H., Hemingway, A., Nier, A. O., and Stuckwisch, C. G. 1941. Reliability of Reactions Used To Locate Assimilated Carbon in Propionic Acid. J. Amer. Chem. Soc., 63, 2140–2.

APPENDIXES

APPENDIX I. GLASS-BLOWING EXERCISE*

OBJECTIVES

These demonstrations and exercises will provide the opportunity to acquire some skill in the simpler manipulations with glass and to gain an appreciation of potentials and limitations of the professional glass blower.

PRINCIPAL EQUIPMENT AND SUPPLIES

Blast burners for gas and air or oxygen mixtures. Both a table model and a hand torch should be available.
Annealing burner
Dydimium glass goggles
Selection of glass-shaping tools and carbon rods
Assortment of glass tubing and rods (Pyrex glass only)
Oxygen cylinder and pressure regulator
Assortment of cork stoppers
Rubber hose

---

* The glass-blowing exercise is presented in the form of an experiment. The authors have found this exercise of such value that they include it among the required experiments in one of their courses.

Ampule files
Roll of asbestos tape
Tungsten wire

PROCEDURE

A. Three demonstrations of glass blowing will be given by an experienced glass blower. These will include the operations listed:

1. Join pieces of 10−12 mm tubing (butt seal).
2. Make a T tube from 8−10 mm tubing.
3. Make right-angled bends in 8−20 mm tubes.
4. Close off tubes and round the ends (test tube ends).
5. Close off tubes and make the ends flat.
6. Draw down tubes and seal to smaller tubes.
7. Join capillary tubes to larger tubes.
8. Join a side-arm of 6−8 mm tubing to a test tube.
9. Join a side-arm to a large beaker or similar vessel.
10. Blow a bulb at the end of a tube.
11. Blow a bulb in the center of a piece of tubing.
12. Make ring seals.
13. Make indentations and small holes in the side of large tubes.
14. Shrink tubes to make a short capillary constriction.
15. Make and seal ampules.
16. Join tungsten to glass.
17. Make a closed tubular circuit.
18. Wrap tubular coils.

These and other techniques are described in the references at the end of this experiment.

B. Each student will be required to perform the operations numbered 1 through 4 of the above list to the satisfaction of the instructor. Conserve the oxygen supply and always turn off the oxygen when the flame is not in actual use. Clean up the glass-blowing table when you are through and be certain that the gas and oxygen are turned off at the source.

C. The student will be required to demonstrate further skill in the manipulation of glass by the construction of one or more of the pieces of apparatus listed:

1. Soxhlet extractor
2. Jacketed condenser
3. Trap or gas-scrubber with a ring seal
4. Glass water-aspirator
5. Chromatography spray-bottle with a ring seal
6. Vigreaux column
7. Constant-level manometer

REPORT

The student will demonstrate to the instructor his skill in each of the four operations in part B and the finished apparatus selected from the list in part C.

QUESTIONS

    1. How can one distinguish soda glass from Pyrex?

    2. How may the following be sealed to a Pyrex tube: (a) tungsten wire, (b) soda glass, (c) platinum wire?

    3. Tabulate the steps in the fabrication of an all-glass, bulb condenser.

References

1. Wright, R. H. 1943. Manual of Laboratory Glass Blowing. Chemical Publishing Co. New York.

2. Nokes, M. C. 1950. Modern Glass Working, 3rd ed. W. Heinemann. London.

APPENDIX II.   FOLIN-CIOCALTEU METHOD FOR PROTEIN (1)

REAGENTS

A. Add 1 ml of 2.7% sodium-potassium tartrate·4H$_2$O and then 1 ml of 1% CuSO$_4$·5H$_2$O to 100 ml of 2% Na$_2$CO$_3$.

B. Commercial Folin-Ciocalteu reagent is diluted with enough water so that it contains 1 $\underline{N}$ acid (as determined by titration to a phenolphthalein end-point).

PROCEDURE

Prepare samples containing between 30 and 200 μg of protein in 1 ml of 0.5 $\underline{N}$ NaOH. Mix 5 ml of reagent A with the sample, and let it stand at least $\overline{10}$ min at room temperature. Add 0.5 ml of reagent B rapidly and mix within 1 sec. After at least 30 min, determine the absorption at 750 mμ in a colorimeter. Determine the amount of protein from an experimentally obtained standard curve. The same quantities of various proteins give different readings with this reagent because the color depends in part on the aromatic residues. The standard curve should be prepared with samples of known concentration of the protein of interest.

Reference

1. Lowry, O. H., Rosebrough, N. J., Farr, A. L., and Randall, R. J. 1951. Protein Measurement with the Folin Phenol Reagent. J. Biol. Chem., $\underline{193}$, 265-75.

APPENDIX III.   FISKE-SUBBAROW DETERMINATION OF PHOSPHATE (1)

REAGENTS

$10\,N\,H_2SO_4$

Reducing agent: Grind 0.5 g of 1-amino-2-naphthol-4-sulfonic acid in 195 ml of 15% $NaHSO_3$ and add 5 ml of 20% Na_2SO_3. Store overnight, or warm until all material dissolves. The reagent is stable for several weeks if kept in a dark, stoppered bottle. Alternatively, grind 0.5 g of 1-amino-2-naphthol-4-sulfonic acid with 30 g of $NaHSO_3$ and 1 g of Na_2SO_3. Store the mixture in a tightly stoppered bottle, and dissolve 7.8 g in 50 ml of water before use.

2.5% ammonium molybdate

30% H_2O_2

PROCEDURE

1. *Inorganic Orthophosphate (and labile organic phosphates such as phosphocreatine)*

The sample (2 to 20 μg of phosphorus) in a volume of 4.2 ml is mixed with 0.2 ml of $10\,N\,H_2SO_4$, 0.4 ml of 2.5% ammonium molybdate, and 0.2 ml of reducing agent, in that order. After 10 min the color is read in a colorimeter at 660 mμ. This method of phosphate determination is sensitive to the pH of the solution and to silicic acid. Alternative methods have been summarized recently (2).

2. *Total Phosphorus*

The sample (2 to 20 μg phosphorus) is digested 30 to 60 min with 0.2 ml of $10\,N\,H_2SO_4$ in an oven, oil bath, or sand bath at $130-160°$. It is removed, cooled, $1-2$ drops of 30% H_2O_2 are added and it is heated to $130-160°$ again for 20 min, then cooled and 0.5 ml of water is added. The sample is heated again at $100°$ for 10 min to decompose pyrophosphate. Now 3.8 ml of water is added to bring the volume to 4.4 ml and inorganic phosphate is determined, as above, without further addition of acid.

References

1. Umbreit, W. W., Burris, R. H., and Stauffer, J. F. 1957. Manometric Techniques, 3rd ed., Chap. 16. Burgess Publishing Co. Minneapolis, Minn.
2. Lindberg, O., and Ernster, L. 1956. Determination of Organic Phosphorus Compounds by Phosphate Analysis. In Methods of Biochemical Analysis, III, 1−22. D. Glick, editor. Interscience Publishers. New York.

APPENDIX IV.   COMPOUNDS SUITABLE FOR BUFFERS IN THE RANGE pH 6 – 8.5

Compound	pK_a
Histidine	6.1
Cacodylic acid	6.1
Di-sodium phosphoglyceric acid	6.2
Sodium bicarbonate	6.3
Di-sodium pyrophosphate	6.5
Sodium phosphite	6.5
Mono-sodium maleate	6.6
Mono-sodium phosphate	6.8
Imidazole	6.8
2,4,6-Collidine	7.4
Sodium diethyl barbiturate (Veronal)	7.7
Triethylamine	7.8
Glycylglycine	8.0
Tris(hydroxymethyl)aminomethane	8.1
Alanine ethyl ester	8.1
Tri-sodium pyrophosphate	8.4

An extensive list of buffers is given in ref. (1). Exact directions for preparing useful buffers over a wide pH range are given in ref. (2).

References

1. Umbreit, W. W., Burris, R. H., and Stauffer, J. F. 1957. Manometric Techniques, 3rd ed., Chap. 9. Burgess Publishing Co. Minneapolis, Minn.
2. Gomori, G. 1955. Preparation of Buffers for Use in Enzyme Studies. In Methods in Enzymology, I, 138–46. S. P. Colowick and N. O. Kaplan, editors. Academic Press. New York.

APPENDIX V.   NOTES TO INSTRUCTORS

The experiments presented in this book were originally chosen for two one-semester courses designed for students who had completed a basic course in biochemistry. Some remarks concerning the organization of the courses are submitted here in the hope that they may be helpful to anyone teaching similar courses.

As given at the University of California, the first-semester course is devoted to the experiments with enzymes (Section II). A one-hour lecture on techniques and seven hours of laboratory work each week are required. The second semester deals with methods of identification and separation (Section I) and radioactive tracer techniques (Section III); this course requires nine hours of laboratory per week and one hour of lecture. Time outside of class is required in performance of a number of these experiments for two reasons. First, materials often have to be prepared before class time; for example, bacteria must be grown for several hours before they are harvested. Second, some experiments such as enzyme isolations cannot be conveniently halted at the end of a laboratory period and must be carried on to a safe stopping point. However, it is undesirable to allow students to devote an excessive amount of time to the course. In particular, for reasons of safety, students should not be allowed to work alone in the laboratory.

In the enzyme course (Section II), Experiments 14 through 22 and 24 through 29 are required of most students. Students who have had experience with an assigned experiment, or who are particularly interested in learning a specific technique are encouraged to select an elective experiment either from this book or from other sources.

In the second-semester course, all students are required to perform the glass-blowing exercise (Appendix I) and Experiments 1, 3, 5, 10, 31, 32, 33, 34, 35, and 36, in that order; in addition, time is set aside for two elective experiments that are to be chosen from other experiments in this book. These elective experiments are either more difficult technically than the required experiments or introduce new techniques. Required experiments are described in considerable detail in order that the student may have the procedures available while he works in the laboratory. Wherever possible, elective experiments are treated more tersely, and only the outlines of procedures are given; the student is presented with a challenge to devise experiments and to refer to the literature for details of procedure.

It is most important that reagents, supplies, and solutions be available; the student can waste a large part of the laboratory time if he must make up solutions by routine, uninstructive procedures. A teaching assistant whose duty, in part, is to handle such work is very helpful.

The list of major equipment used in these courses may seem rather extensive, and the instruments are certainly expensive. However, these instruments are essential in modern-day biochemical research, and it is necessary that students become familiar with them. Since ten to twenty students are commonly enrolled in these courses each term at the University of California a number of students may simultaneously need the same piece of apparatus, and some means to prevent overcrowding is necessary. It is suggested that, if necessary, the experiments be rotated through the class, rather than that groups of students work together. In our experience, it has proven best to allow students to work individually, although a pair of students of equal ability can gain much from one another and can accomplish more than each can alone. Groups of three or more students have not worked well.

Lecture periods in each of these courses are, in the opinion of the authors, of great importance. In these lectures, certain aspects of theory, the range of application, and a critical evaluation of the techniques are presented. In order to hold this book to a modest size, only the outline of the material in these lectures is presented here, but references are listed which treat the various subjects more extensively. At the University of California the general theory of enzymes is presented in a separate lecture course which covers the material given in Outlines of Enzyme Chemistry by J. B. Neilands and P. K. Stumpf (John Wiley and Sons, New York).

APPENDIX VI. LIST OF MAJOR EQUIPMENT

The list below represents major equipment for the experiments required at the University of California.

Section I

1 Vacuum distillation column
1 Beckman model DU spectrophotometer
1 Mechanical vacuum pump
1 Oven (to 150°)
1 Craig countercurrent distribution machine
1 Fraction collector
1 pH meter
1 Abbé refractometer

Section II

1 pH meter
1 Clinical centrifuge
1 Refrigerated centrifuge
1 High-speed centrifuge
1 Beckman model DU spectrophotometer
1 Warburg respirometer
1 Constant temperature water bath (25°–60°)
1 Colorimeter
1 Ball mill
1 Autoclave

Section III

1 Autoclave
1 Geiger-Müller end-window tube and counter
1 Clinical centrifuge
1 Constant temperature water bath (25°–100°)
1 Sample spinner and dryer

APPENDIX VII. LOCKER INVENTORY

A list of equipment needed for all major experiments described in this book is presented below.

```
12  Beakers, Griffin, 50 ml to 2000 ml
 2  Bottles, dropping
 1  Bottle, polyethylene, 500 ml
 1  Bottle, reagent, 500 ml
 1  Bottle, reagent, 2000 ml
 1  Brush, test-tube
 1  Burette, glasscock, 10 ml
 1  Burette, glasscock, 50 ml
 1  Burner, Bunsen

 2  Clamps, burette, pressed steel
 2  Clamps, holder
 1  Clamp, pinchcock, Day
 1  Clamp, screw
 1  Clamp, test-tube
 1  Cylinder, graduated, 10 ml
 1  Cylinder, graduated, 100 ml
 1  Cylinder, graduated, 250 ml
 1  Cylinder, graduated, with glass stopper, 100 ml

 2  Dishes, evaporating, 70 ml
 2  Dishes, evaporating, 200 ml

 1  Flask, boiling, round-bottom, 200 ml
 1  Flask, boiling, round-bottom, 500 ml
18  Flasks, Erlenmeyer, 50 ml to 1000 ml
 1  Flask, filter, 500 ml
 7  Flasks, volumetric, Pyrex, with glass stopper,
        5 ml to 1000 ml
 1  Forceps, bent tip
 1  Funnel, Buchner, 91 mm
 3  Funnels, long stem, 5 in.
 1  Funnel, separatory, 125 ml
 1  Funnel, separatory, 500 ml

 1  Lighter

12  Pipets, serological, 0.1 to 10.0 ml
11  Pipets, volumetric, 1 ml to 25 ml
 6  Planchets, counter, aluminum
 1  Plate, spot
```

    1 Rack, wire test-tube
    1 Ring, iron, 3 in.
    1 Ring, iron, 5 in.
    1 Ring, support, cork, 3 in.
    1 Ring, support, cork, 5 in.
    1 Rod, Pyrex, 3 ft

    1 Scoopula
    1 Spatula, double-end
    1 Spatula, 8 in.
    1 Support, iron, small base, 20 in. rod
    1 Support, iron, large base, 24 in. rod
    1 Support, test tube, wood

    1 Thermometer
    1 Tripod, 6 x 9
    1 Tube, $CaCl_2$
    2 Tubes, calibrated, Klett
    6 Tubes, centrifuge, plain, 12 ml
    3 Tubes, centrifuge, plain, 40 ml
   12 Tubes, test, 10 x 75
   24 Tubes, test, 15 x 125 without rim
   14 Tubes, test, 20 x 150
    6 Tubes, test, 25 x 150
       Tubing, red rubber, 10 ft
       Tubing, vacuum, 5 ft

    6 Watch glasses, 3 in. to 6 in.

Miscellaneous items: file ampoule, pH papers, pipe cleaners, razor
                      blades, sponge, wax pencil, wire gauze.

INDEX

185